T0220279

Testing Data Warehouse Applications:
Beauty Will Prevail

Written By: Doug Vucevic & Mengxi Jett Zhang

Edited By Ned Lecic

Cover Design By Jana Jelovac

ISBN: 978-1-4269-9389-3 (sc)

Trafford rev. 08/27/2011

www.trafford.com

North America & international
toll-free: 1 888 232 4444 (USA & Canada)
phone: 250 383 6864 ♦ fax: 812 355 4082

This book is dedicated to my wife Ksenija, who makes everything possible and our children: Drasko, Diana and Vukan, You guys may not know this, but you inspired me to make this book happen.

Doug

I dedicate this thesis to my parents, wife, and children. Without their support, understanding, patience, and most of all love, the completion of this work would not have been possible.

Jett

Acknowledgment

First and foremost we the authors are grateful to Quest Software for providing tools and permission to use the screen prints taken with: Quest® TOAD®, and Quest® Code Tester for Oracle. Our gratitude also goes to HP Company for permission to use the screen prints taken with their tools: HP Quality Center® and HP Quick Test Professional [QTP]®

The authors want to extend a sincere thank-you to Draga Dragasevic, her guidance and thorough review of the good part of the manuscript has greatly improved the clarity of the book. Thanks to Dr. Vladimir Pantic for reviewing the introductory section on DW.

We have been blessed with the great fortune of working at some of the most excellent Canadian companies like, IBM, Canadian banks like, Bank of Montreal [BMO], Toronto Dominion [TD], Canadian Imperial Bank of Commerce [CIBC], Royal Bank [RBC], Government of Ontario, Canadian retailers, Hudson Bay and Canadian Tire, Canadian aircraft manufacture, Canadair, etc. We are most thankful to these great organizations as without the knowledge and the experience that we have acquired at these organizations the book in this form would have not be possible. In particular, we are grateful, for sharing their thoughts and suggestions, to the exceptional QA managers, Scott Coolling of BMO, David Wu of CIBC and Wilson MacArthur ex-IBM manger and Steve Labrecque of Canadian Tire Corporation.

Table of Contents

Acknowledgment

Overview --- 1

Introduction --- 3

Introduction to Data Warehouse Application ----------------------------- 5

    Data Warehouse -- 5

    ETL – Extract Transform Load -- 11

    BI - Business Intelligence --- 14

    Data Mining -- 17

Missing Link in BI Success -- 19

Quality Assurance Story --- 31

    Software Testing vs. Software QA ----------------------------------- 38

    Validation vs. Verification -- 39

    What's different? -- 42

Quality Assurance Strategy for DW Application ----------------------- 47

    Validate, Validate, Validate! --- 50

    Data Visualization -- 51

    "War and Peace" by Tolstoy presented visually -------------------- 60

    Seeking Relationships -- 63

Testing – Mostly about Verification -------------------------------------- 69

    Data Quality --- 70

    Unit testing -- 73

    Unit Test Automation -- 76

    System Testing -- 85

Data Completeness --- 86

Data Transformation --- 87

Enterprise Integration Testing of DW ---------------------------------- 90

Regression testing --- 96

Performance testing --- 97

User Acceptance Testing -- 99

Risk Management --- 101

Strategy Review Time ---108

Power of storytelling --108

Storytelling in Business ---112

Journey through Bermuda Triangle --------------------------------------120

It is much later than you think! ---123

Testing - Moving from "What" to "How" ------------------------------- 127

Test Automation Case Study -- 137

Test Case 1 --- 137

Test Case 2 --- 158

Test Case 3 --- 167

Test Case 4 --- 170

Test Management --- 181

References --- 185

Overview

> *"In this new world, information is king. The more information you have, and the better and faster your analysis, the greater the probability that you will make winning investments".*
>
> ~ Geoffrey More "Living on the Fault Line"

Testing Data Warehouse Application is a practical guide in testing and assuring Data Warehouse [DW] applications. It first appeared in the form of handouts that we gave to our students for a course we teach at the Institute for Software Engineering®. It grew out of our frustration while trying in vain to find the appropriate reference material for the Data Warehouse testing course. We marshalled our own resources and you are reading the result of it. The book is not based on rigorous scientific evidence. Rather, it is a tale from the trenches of testing battlefields, a message passed from warrior to warrior.

A data warehouse is a valuable corporate asset used to envisage business strategies and make informed business decisions. The enhanced access to information that a data warehouse provides enables an organization to make time-critical business decisions that are required to remain competitive. Data warehousing needs a comprehensive assessment of the impact to the entire organization and development of a plan for an organized, systematic solution. As for the Quality Assurance [QA] teams, it creates an exciting new opportunity that comes once in a lifetime. It is nothing less than a new business paradigm which creates a new unlimited learning opportunity [essential if one wishes to prosper in it]. As with any new paradigm, most of us are unprepared for it. That is bad news. The good news is so is: so is everyone else. The race is on! The most nimble of us will flourish the most. Read on! This book will reword you with a head start.

Enterprise Data Warehouse [EDW] is a mission-critical asset because it feeds important business intelligence applications used in making strategic business decisions, such as business performance optimisation, revenue enrichment, customer service, etc. Defects in EDW not only increase the cost associated with rework, but also result in lost business opportunities that cannot be known, thus cannot be accounted for or, recouped. In view of this, we strive to walk the reader through the testing and quality assurance activities required to minimize the risk of production problems caused by the erroneous use of data. If we are successful with this book, your goal of delivering near problem-free DW applications will be achieved more easily.

Business knowledge, acquired from EDW, is a result of transforming data into information and finally into Business Intelligence [BI]. The goal of this book is to show an actionable QA methodology and practical testing techniques for delivering near problem-free DW applications to our organizations. Ours [QA] is the responsibility of ensuring that this technology helps our organizations in maximizing business opportunities by helping them make better decisions and ultimately giving our customers a more rewarding experience. QA professionals must always keep in perspective that DW application is a solution to a business problem and if the business problem is not solved for whatever reason, be it incorrect business requirements, wrong design, or coding errors, then the product does not deliver the business benefits it is designed to deliver.

Ours is an era of the global marketplace and the new differentiator in that marketplace is the effective deployment of decision support technology. EDW is an enabling intelligence-driven technology. An effectively implemented DW application can provide a full picture of a business and give insight into the future risks it faces. We are at the historic junction of horse-versus-locomotive competition. Those who capitalize on this new opportunity will emerge as future market leaders.
It is said that the "Data is the New Oil", but data alone is not enough, it is our ability to create the business knowledge based on that data.

Yes, we are delivering a message of warning, except that our message is not accompanied with despair, but with the hope of a brighter future for all humanity at this critical junction with the new paradigm. We are framing this message into the larger context and relating it to a journey through unknown lands and the stormy seas. Journey that we all have to take, that journey is called life.

Introduction

These are changing times where only a third of excellent organizations continue to maintain that status over the long term, where even fewer are able to implement successful change programs, where everything is going so much faster and where skills once considered essential become obsolete, even worthless and more quickly now than ever before. Trouble is we don't know which skills will be in most demand tomorrow. This book attempts to identify some skills that will be in demand in the immediate future, but perhaps more importantly there are some timeless qualities, a mindset that will always be in demand. Every employer will always need people with right mindset - persistent, energetic, innovative, optimistic and resilient. This right mindset will always win against any skill set.

Therefore the book is not a pure technical book, rather it is a technical book framed in a larger context. On the higher level the book is comprised of three main themes that span across all chapters:

1. Motivation for writing this book,
2. Description and causes of the predicament we are in and its impact on society and each one of us.
3. Exploring solutions to the problem and the opportunity it is creating.

An introductory chapter on the DW concepts and its components provides basic explanation of the software we are about to start testing. Good references are provided to the QA professionals interested in pursuing career in this [DW] fastest growing field of Information Technology [IT]. For a better understanding of Data Warehousing we strongly recommend an excellent introductory books by Ralph Kimball, Margy Ross *"The Data Warehouse Toolkit: The Complete Guide to Dimensional Modeling"*[Ref. 1] and by Inmon, W.H.: *"Building the data warehouse"*[Ref. 4].

The other rationale for the introductory chapter to DW section is to show that this knowledge is a prerequisite [as it is the case in testing any software] for the Quality Assurance [QA] teams that intend to pursue DW application QA assurance and testing.

Introduction of Business Intelligence, as this is the main challenge in testing and constricting DW Application. Major obstacles and remedies are discussed in section "Missing Link in BI Success".

After a brief introduction to Data Warehousing technology, in the next chapter of the book QA processes and methodology in general are discussed. Differences between software application and DW applications are considered. Specific strategy for testing DW applications is recommended. This chapter describes methodology to deliver near problem-free DW applications into production. Here we included a discussion on the Data Warehousing application testing cycle and how it relates to "Software Development Life Cycle" [SDLC].

Next chapter contains the description of the test environment and test tools we are using in demonstrating the concepts presented in previous chapters of this book. Description of how to develop and build a DW application test environment and how to deploy the appropriate testing tools is included.

Even though, the major part of the book is devoted to what and how to do it, considerable part the book is devoted to "why". The great lives in human history have been built on "why". If the person knows why, she or he will learn how, despite all the obstacles. The key in achieving anything is not how, but why. Section on Risk and Review of QA Strategy devote considerable amount of time discussing motivation and positioning DW application correctly within an organization.

The last section of the book is where demonstrate hands-on examples to illustrate how to find, develop and execute the test cases with the focus on various testing techniques that may be employed in testing EDW. The case study for concept developed and demonstrated with an example of using automation test tools for regression testing. Software tools like, Quest TOAD®, HP Quality Center®, and Quick Test Professional [QTP]® used to illustrate typical real life environment. We also present "end-to-end" user test form data sources to [BI], testing correctness of the Reporting and the Analysis tool.

Introduction to Data Warehouse Application

Data Warehouse

A Data Warehouse [DW] is a database system in which data is collected in order to be analyzed. Enterprise Data Warehouse [EDW] is an information environment, a new paradigm with the specific intention of providing vital strategic information. Most of a company's data is collected in order to handle the company's ongoing business. This is called "operational data" and includes categories such as CRM [customer relationship management] systems, SCM [supply chain management] and databases containing various transactions. The system from which data is collected contains the operational data; hence it is referred to as OLTP [Online Transaction Processing]. The Decision Support System [DSS or DW] provides a good physical separation from its OLTP. DW is a tool that integrates an organization's historical and heterogeneous data into an information source which enables "Online Analytical Processing" [OLAP].

The term Data Warehouse actually refers to a collection of relevant data from multiple sources that is rationalized, summarized and catalogued in stable, long-term data storage, facilitating the management's decision making process. Major characteristics of DW are:

- *Subject -oriented* – data that provides information about a particular subject instead of a company's ongoing operations.
- *Time -variant* – all data in a DW is identified with a particular time period.
- *Integrated* - data is gathered from various sources and merged into a coherent whole.
- *Non-volatile* – data is never destroyed.

The subject-oriented data of a DW is organized around the functions of the organization. Information in a DW is organized into various *"dimensions"*. For example, for the retail company in our case study at the end of this book, major subject areas [dimensions] might be: products, orders, vendors, sales, customers, etc. A sales analysis database is organized according to products, time, territory, and other dimensions. An invoice database could use time, customer, product, and supplier dimensions. Each type of company has its own unique set of subjects.

In practical terms, a Data Warehouse is a collection of technologies that enable business users[1], such as financial experts, planners, executives and various analysts, to make faster and better strategic decisions. Data warehousing products encompass numerous suites of products for analysis, query and reporting capabilities that also include DSS characteristics. They consist of software applications that maintain the database and present the data. Metadata [data that describes the data] is usually is stored in a Metadata Repository, but it is also part of every DW. The purpose of the EDW is to provide the business value of the data to organizations. The business value of the EDW is directly relevant to data accuracy and information quality, which will be addressed later in this book. DW is a de-normalized, business-centric *star schema* - a standard data warehouse design with one or more *fact* tables comprising the hub of each star, surrounded by various *dimension* tables that allow the level of granularity of the *facts* to be drilled into or rolled up along relevant vectors. When we are talking about DW we are most commonly referring to the Star-Join Schema, which consists of:

- *Fact Tables* – "what are we measuring?"
- *Dimension Tables* - "what are we measuring with?"

This is quite different from the fully normalized relational architecture adopted by most OLTP systems. The *star schema* does not take data redundancy into consideration as quick response time is the highest priority. The schema is de-normalized; therefore, accessing and filtering huge data sets is extremely fast, since table joins are kept to a minimum and data access paths are fully pre-planned in accordance with current business needs.

Robust indexing, which normally has an impact on performance during *insert, delete* or *update* transactions does not affect performance, since only *select* transactions by end-users are allowed. This also eliminates the kinds of *insert, delete* and *update* anomalies inherent to de-normalized schemas.

According to computer scientist Bill Inmon's Time-Variant concept, every unit of data in the DW is accurate at some point in time. In many cases, a record is time-stamped, or it has a transaction date. In every case, it has some kind of time marking to show the time when the record

---

[1] The person using DW application – person who has to interact with the software

was accurate. It calls for storage of multiple copies of the underlying detail in aggregations of differing periodicity and/or timeframes. DW may have details for seven years along with weekly, monthly and quarterly aggregates of differing duration. The time variant strategy is pivotal, not only for performance but also for maintaining the consistency of reported summaries.

Integrated data is the most important aspect of the DW. Integration means that data from a variety of sources is not commingled [blended together]. Integration is the process of mapping dissimilar codes to a common base, developing consistent data element presentations and delivering this standardized data as broadly as possible. The DW is fed from multiple and disparate sources of data. Data is then transformed and reformatted into a single corporate image that can be used by decision makers on the corporate level.

Non-volatility of data literally means that once a row is written, it is never modified. This is necessary to preserve incremental net change history. This, in turn, is required to represent data as of any point in time. When you update a data row, you destroy information. You can never recreate a fact or total that included the unmodified data. Maintaining "institutional memory" is one of the higher goals of data warehousing. Data in the operational database is regularly accessed and modified one record at a time. Data in the DW is usually mass-loaded but it is not changed.

A basic assumption of Inmon's is that a data warehouse is exclusively a store of data for decision support. Under his definition, this precludes the use of a data warehouse for what he calls the Operational Reporting Process[2]. In recent years, DW has become the main trend in corporate computing, as it provides managers with the most accurate and relevant information to improve strategic decisions.

Planning and analysis applications called On-Line Analytical Processing [OLAP] are the heart of a data warehouse. The term "MOLAP" for Multi-Dimensional OLAP is also used because, unlike OLTP entity-relationship models consisting of two-dimensional tables, data warehouses use a *"multi-dimensional"* model for storing data. The term

---

[2] INMON, W.H.: 'Building the data warehouse' John Wiley and Sons, September 1998,41, 9, pp. 52-60 New York, 1997, 2nd Edition

OLAP is neither a meaningful definition nor a description of what an OLAP is.

An OLAP is a series of protocols used mainly for business reporting. With OLAP, businesses can analyze data in all manner of ways, including budgeting, planning, simulation, DW reporting and trend analysis.

OLAP is a type of software technology that gives a company's various managers and executives a capability for quick and interactive analysis of multidimensional information – *"derived"* data which has been transformed from raw *"primitive"* data. Below are some differences between *derived* and *primitive* data based on W. Inmon's recommendations:

- Operational data is primitive, whereas DSS data is derived.
- Primitive data is detailed data used to run day-to-day operations. Derived data is summarized for the DSS needs of the company.
- Primitive data can be modified. Derived data can be recalculated but cannot be changed.
- Primitive data is current value data. Derived data is historical data.
- Primitive data is operated on by repetitive procedure. Derived data is operated on heuristic programs and procedure.
- Primitive data supports the clerical function. Derived data supports the management decision making function.

Operational databases have a high number of transactions that take place every hour. This kind of database is always up-to-date and represents a snapshot of the current business state at a given point in time. DW, on the other hand, is an informational database and is static and stable. DW is refreshed, usually nightly, and represents a snapshot of the business at a given point in time in the past. This is why DW is not used to make operational, but strategic decisions, which are based on events in the past.

When we are talking about DW, we are most commonly referring to the Star-Join Schema, which consists of:

- *Fact Tables* – "what are we measuring?"
- *Dimension Tables* - "what are we measuring with?"

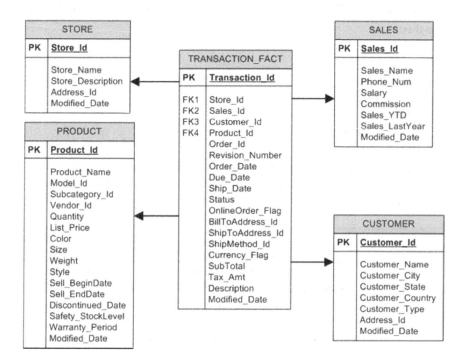

Fig 3.1 Star-Join Schema

The Fact Table is at the center of the data model. Fact tables contain keys to all the dimension tables and measurable facts required by data analysts. For example, a store selling computer parts might have a fact table recording the sale of each item. Measurable numeric facts [numeric attributes such as Sales Dollars, Invoice Amount, etc.] are stored in a fact table. Fact tables can grow very large, with millions or even billions of rows. It is important to identify the lowest level of facts that it makes sense to analyze for the business; this is often referred to as the fact table "grain" which determines the size of the fact table. To access multi-dimensional information, the fact table also contains all *foreign keys,* which are used to join the facts to all *dimension tables.*

Dimension tables organize and index the data that is stored in a fact table. A visual representation of this connection between fact tables and dimension tables appears as a star, hence the name. Dimension tables are typically small, ranging from a few to several thousand rows. However, dimensions can occasionally grow large: a large bank could have a customer dimension with millions of rows. Dimension table structure is typically very shallow; for example, the customer dimension could look like this:

Customer_ID
Customer_name
Customer_city
Customer_state
Customer_country

Summarization, or creating rollup summary tables, can be set up with a "materialized view". These summary tables help avoid heavy queries on the fact table.

Examples of the most common basis for summarization and relation to the fact and dimension tables	
Measures:	Allow changes to the view of data, such as Periodic, Week To Date [WTD], Month To Date [MTD], Quarter To Date [QTD], Year To Date [YTD]. Daily transactions are aggregated to provide consolidated weekly or monthly comparisons viewed using a Calendar interface
Dimensions:	I.E. Services or Products and their properties [COLOR or SIZE] can be totaled into a hierarchy of BRAND, MANUFACTURER, CATEGORY, or other aggregate.
Currency:	"Local Currency" and/or "Reporting Currency"

Every dimension, such as time, can be structured into a hierarchy of consolidation levels - years, quarters, months, weeks, individual days, or other levels of data granularity. But "day of the week" is an extended attribute.

The lower [finer, more detailed] the level of granularity available for analysis, the more costly it is to store and process the data.

One other important aspect of EDW is its very large size. For example, Wal-Mart's data warehouse stores 570 terabytes. A terabyte is equivalent to 1024 gigabytes [the prefix "tera-" is derived from the Greek word for a monster]. Google data warehouse stores about 4 petabytes. A petabyte is equivalent to 1024 terabytes [the metric prefix "peta-" means 10^{15}].

A DW is planned and produced to support the decision-making process in an organization. Data are obtained from the production databases and reorganized in the DW, so that queries can be efficiently answered, without hindering the performance and consistency of the production systems. DW is a tool from where businesses, using applications, obtains the data to provide the information that will contribute to the business' success. Data warehouses are optimized for speed of data retrieval. The data in data warehouses are de-normalized as a consequence of a dimension-based-model. This is necessitated to satisfy speed of data retrieval. Data in DW are stored multiple times – in their most granular form and in summarized forms called aggregates.

ETL – Extract Transform Load

ETL stands for "extract, transform and load". It incorporates the dispersed data from all departments within an organization into a uniform DW. Most often, ETL systems stage the data once or twice between the source and the data warehouse target. In our example at the end of the book, we have one stage and in this case, the stage is the database [often, the staging area is in the file system].

Extraction is the process of moving operational data into the DW. Data from "online transaction processing" [OLTP] and legacy systems provide iinflow into the staging servers of a data warehouse. In order to build the data warehouse, the appropriate data must be located. Typically, this will involve the current OLTP system, where the "day-to-day" information about the business resides, and historical data for prior periods, which may be contained in some form of "legacy" system. Often, these legacy systems are not relational databases; much effort is required to extract the appropriate data. For example, in a bank, data may be gathered from loan processing, pass book processing and accounting systems. In a retail store, data may be gathered from point-of-sale devices and cash registers. Operational data can be in the form of flat files, table records of the relational database, Excel files, etc. The extract program rummages through a filing or database system using specific criteria for selecting or filtering qualified data for extraction and transformation into the other filing or database systems. In order to find qualified data, many files and formats have to be analyzed. Some of these files are stored in an Information Management System [IMS]; others use the Virtual Storage Access method [VSAM], while some use the Integrated Data Management System [IDSAM]. Diverse skills are required for assessing and testing in these environments. An additional

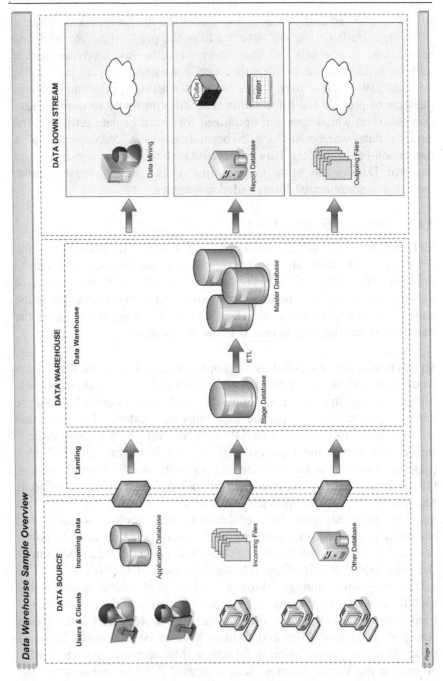

Fig. 3.2 ETL Environment

complication can be the fact that the element may be stored in two or more of these files under a different name. For example, the entity CUSTOMER may be stored in another collection of data in which the same entity is known under the name of CLIENT. Every piece of data must be analyzed and "rationalized"; otherwise, we will be mixing "apples and oranges" in our reports.

An advantage achieved by the extracting process is that data is copied away from the high-performance operational on-line processing environment so it does not affect the performance of that environment.

Transformation is carried out after the DW schema has been designed. It is a process of changing the data from being transaction-suitable to a structure that is most suitable for decision support analysis.

Data extraction includes a range of data grooming actions which are performed to make data presentable for DW. Typically, this is performed through a process known as "Data Cleansing", usually in conjunction with the data transformation phase of ETL. A data warehouse that contains incorrect data is not only useless, but also very dangerous. The whole idea behind a data warehouse is to enable decision making. If high-level decisions are made based on incorrect data in the warehouse, the company could suffer severe consequences. Data cleansing is a process that validates and corrects the data before it is inserted into the warehouse. For example, the company could have three "Customer Name" entries in its various source systems, one entered as "RBC", one as "R.B.C." and one as "Royal Bank of Canada". Obviously, these refer to the same customer. A business decision must be made as to which is correct, and then the data cleansing tool will change the others to match the business rule. The transformation process includes standardizing, integrating, cleansing, augmenting, aggregating and creating the data sets for loading into the repository. This is why data transformation is the most complex and biggest challenge in building DW.

Load deals with loading the DW with the previously transformed data. The DW has to be populated automatically [usually nightly] and consistently to reflect changes on the operational systems. Loading is a straightforward process. The key consideration in the loading process is to achieve the appropriate speed of data loading. This is achieved by using various methods, such as:

- Pre-sorting the file as per the primary key index for Data Warehouse loading. This is the most highly recommended technique.
- Turning off Logging during loading.
- Dropping and recreating Indexes. This, as well as turning off Logging, can save on the overhead during the loading.

BI - Business Intelligence

"Water, water everywhere, nor any drop to drink..."

~ Samuel Taylor Coleridge

These famous lines by Samuel Taylor Coleridge from *The Rime of the Ancient Mariner* are reminiscent of the state of BI in most organizations today. Despite an abundance of data, the staff is often thirsty for information. We are in the midst of an "information explosion" in which there is too much information for anyone to absorb and analyze. The amount of data available has increased dramatically in the last few years, but the ability to make sense of it has increased little, if at all.

> *Our networks are awash in data. A little of it is information. A smidgen of this shows up as knowledge. Combined with ideas, some of that is actually useful. Mix in experience, context, compassion, discipline, humor, tolerance, and humility, and perhaps knowledge becomes wisdom.*

Turning Numbers into Knowledge, Jonathan G. Koomey, 2001, Analytics Press: Oakland, CA page 5, quoting Clifford Stoll.

Today's companies have years' worth of customers' financial and other data saved in their various operational systems. A corporation's data resides in the different platforms and different operational units of the corporation. The question that begs an answer is: if we have these mountains of data in our organizations, then why can't our executives use the data for strategic decision making? The problem we are facing is - how do we get from data to information? Consider, for example, this problem from the banking industry:

"Compare account activities from this year with account activities from the last 10 years."

In order to arrive at the answer, a DSS analyst must deal with lots of non-integrated applications. For example, a bank may have different applications for loans and mortgages, separate saving and checking applications, and so on. Trying to get information from them on a daily basis is not practical, nor is it even possible. The major obstacle is that the historical data may be unavailable in those systems. For example, deposits may have data for the last 18 months, while the mortgage department has data for the last 25 years, but the loan department may have data for last two years, and so on. The systems in that environment are inadequate for supporting today's information needs. Two major obstacles are the lack of integration and the unavailability of historical data. That is why our corporations today are drowning in data and are facing an *information crisis*.

Business Intelligence [BI] comes to rescue. BI is a process of turning data into information and then into business knowledge. The knowledge obtained by means of it can pertain to customer needs, as well as insight into customer decision making processes, industry trends, general economic trends and conditions, social trends, and other fields of interest. Essentially, a new data architecture had to be developed in order to support the new information requirements. The concept of *derived data* based on operational *primitive data* is required in support of the DSS decision-making process.

The data needed for strategic decision making is the data from all those other systems, integrated and derived into an EDW, before it becomes suitable for any analysis. Therefore, EDW contains data suitable for strategic decision making for the business intelligence of an enterprise. This environment is completely separate from day-to-day operational environments. Only then can we think of applying BI, not as a product or an application, but as an architecture with a collection of integrated databases, decision-support applications and tools that provide the business community with easy access to business knowledge.

An important part of Business Intelligence is *analytics*. Analytics employs extensive use of data visualization, quantitative analysis, statistics, and explanatory and predictive modeling to support a fact-based strategic decision-making process. Business Intelligence includes all the processes involved in data access, reporting and analytics.

Technically analytics could be done using paper and pencils as it's the ideas, not the tools that really count, but in today's state of IT technology and the amount of data there is no need to do it that way. Instead, we should be looking for automated analytical tools. This book explores analytics using Excel, but there is a whole range of other statistical and predictive tools could be used for Business Intelligence. The important point that we are making here is that analytics is not about analytics information software, but the human interpretation of the data underneath it.

Some aspects of data visualization are illustrated in the data visualization section, as this is where the business intelligence industry will need to turn to succeed in the future. In today's business environment, in which our brains are bombarded with much competing information, decision makers don't have time to invest in searching for the information; that is why a goal of the information visualization must be that the information is available at a glance.

Statistical regression not only produces predictions, but the prediction that is accompanied by the confidence level of that prediction. Statistical data analysis provides the most powerful tools for understanding data, but the systems currently available for statistical analysis are based on an outdated computing model and have become much too complex to be used for this purpose. What we need is a simpler way of using these powerful analytical tools, which transparently presents data statistics visually in a much a simpler way. Statistical analysis uncovers hidden relationships among widely disparate kinds of information. The statistical analysis tools present the results of the analysis in a way that helps ensure that our intuitive visual understanding is commensurate with the mathematical statistics hidden under the surface. Thus, visual statistics ease and strengthen the way we understand data and they are used in information visualization of the abstract data underneath to amplify our cognition and create business insights.

Data Mining

> *"The quiet statisticians have changed our world; not by discovering new facts or technical developments, but by changing the ways that we reason, experiment and form our opinions."*
>
> ~ Ian Hacking

There are many definitions for Data Mining [DM], but one from Gartner Group seems to be the most comprehensive:

> *"The process of discovering meaningful new correlations, patterns, and trends by sifting through large amounts of data stored in repositories and by using pattern recognition technologies as well as statistical and mathematical techniques."*

DM is the science of extracting useful information from large data sets or databases. It is also known as Knowledge Discovery in Databases [KDD]. KDD is a multidisciplinary field covering Information Retrieval, Machine Learning Pattern Recognition, Statistics, Artificial Intelligence Expert Systems, Visualization, and Databases. DW is a useful source of data for exploration and data mining; as the data in the DW is cleansed, integrated, categorized and historical, it becomes conducive for exploration and data mining.

Generally, data mining [sometimes called knowledge discovery] is the process of analyzing data from different perspectives, finding patterns and summarizing them as useful business information - information that can be used to increase revenue, cuts costs, or both. Patterns, associations or relationships among all this data can yield information. For example, analysis of retail point-of-sale transaction data can yield information on which products are selling and when. This information can be converted into *knowledge* about historical patterns and future trends. For example, a summary of information on retail supermarket sales can be analyzed in light of promotional efforts to predict the knowledge of consumer buying behavior. This allows the producer or retailer to determine such things as which items are most susceptible to promotional efforts. Stores like Wal-Mart are constantly making an effort to have no excess inventory at all. What they have on the shelf is all they've got.

An important aspect of data mining is data visualization: the visual interpretation of complex relationships in multidimensional data where graphics tools are used to illustrate data relationships.

As we could all testify to the growing gap between the generation of data and out understanding of it. As the volume of data increases inexorably the proportion of people that understand it deceases alarmingly. DM is defined as process of discovering patterns in data. There is nothing new about this. People have always been seeking patterns, hunters are looking for patterns in animal behavior, farmers are looking in weather patterns, lovers seek patterns in their partners' responses, etc. In data mining data is stored in computer memory and searches are automated. What is new is ever growing number of opportunities for finding new patterns. As the amount of date swells and with the abundance of computing power data mining becomes our hope for elucidating hidden pattern in data. Intelligently analyzed data becomes strategic advantage which can lead to new insight bringing about competitive advantage.

The ultimate goal of data mining is prediction and predictive data mining is the most common type of data mining and has the most direct application to business. Predictive analytics could be used to predict how customers would respond to a future promotion. In fraud detection to identify which credit card number may be fraudulent in nature, etc. By learning from the abundant historical data, predictive analytics provides the business value beyond standard business reports and sales forecasts: by providing actionable predictions for each customer.

The idea of learning form data has been around for a long time, but despite of these nice promises so far goal of DM, as most of the gold rashes of the past, has been to "mind the miners". The largest profit has been in selling tools to the miners. DM too has been used as a means to sell computer hardware and software. A quote of Chuck Dickens [ex director of Computing at SLAC] has been:

> *"Every time computing power increases by a factor of ten, we should totally rethink how and what we compute".*

Since all currently DM tools have been invented, computing power and size of data has increased by a several orders of magnitude. So we should soon look forward to much brighter future for the new DM methodology[3].

---

[3] Jerome H. Friedman: Data *Mining and statistics: What is the Connection?*

Missing Link in BI Success

"If you love what you do, you will never work another day in your life."

~ Confucius

Gartner's report released in February 2009 on the state of global BI and based on feedback from 1,400 CEOs clearly stated that companies will continue to invest in BI in spite of the fact that they were not getting the anticipated business value from it. What is more important, all 1,400 CEOs committed to supporting their BI efforts and to making it the number one priority for the 2009 twelve-month period. In fact, BI has been on the top of the CEO's priority list for the third consecutive year, stressing the important role of BI in strategic planning.

BI projects, however, are known for their high failure rate, broken promises on delivering business value, and budget and timeline overruns. Nevertheless, BI is still perceived as one of the remaining initiatives that can make companies more competitive. That is because the potential for enormous returns on investment and the competitive advantage make data warehousing difficult to ignore. It appears at this time that only large, rich companies can tolerate the high cost and very low success rates.

If we accept the popular scapegoat claim that software engineers are to blame for the high failure rate of BI projects, we will never be able to understand the real cause behind this problem. Consequently, if we don't know what the problem is, it will never be rectified.

Software engineers are facing a problem for which there appears to be no solution at this time. The "systems thinking" in our linear "if-then" or "either-or" world obviously does not work for us any longer, as the solution to this problem spans across other systems which we have not been trained to understand and which, therefore, cannot provide a solution. Software engineers often provide product information [such as charts and graphs] based on their intuition, knowing little about the effectiveness of graphic displays and human perceptual capabilities. Designers' focus is typically on finding out which design works best rather than on understanding how and why. Since information visualization is said to support decision making, we propose that information design practice must include knowledge about the domain which enables decision making, namely the brain.

Looking from a different perspective, what is involved in delivering a full BI solution? First, it is necessary to build the data warehouse and then design the application to create graphs and reports which are intended to assist business users in predicting the future. Clearly, these activities fall into two groups: engineering and humanistic, as is demonstrated in the table below:

Engineering activities	Humanistic activities
Extracting, Transforming and Loading [ETL] data	Detecting patterns
Data analysis	Synthesizing and finding meaning
Reporting	Predicting

Predict the future?! Can an engineer or a programmer create objects to detect patterns and business opportunities, produce artistic and emotional beauty, craft a satisfying narrative, which enables connecting unrelated ideas into something new, empathize with clients in order to understand the subtleties of human interaction, find joy in one person's life and elicit it in another's, and stretch thought for purpose of finding meaning in life? Can they provide graphs and charts which are both efficient and sufficient to produce insights leading to better or even outstanding business decisions? Definitely not, at least not at this time! Yet, these are the very objects that are required by business. That is why our current BI solutions have hit the proverbial brick wall. To surpass these challenges, we are now turning to other fields of human endeavor.

In order to explain this phenomenon, we have to take a diversion to explain what brought us here and investigate possible solutions. Fortunately, we do not have to look too far, as both the problem and the solution is in our heads. It is our brains that have brought us into this mess and it is our brains that will take us out of it.

The research of 1981 Nobel Prize-winning neurobiologist, Roger W. Sperry, whose work was first published in 1968, proves that the human brain is comprised of two hemispheres specialized for two fundamentally different ways of thinking. The left side of the brain is convergent - involved with linear [logical and sequential] thought. This is the side we use to solve mathematical problems and construct content for written reports. The right side is divergent - concerned with imagery, such as recognizing patterns [e.g. a face in a crowd], intuition, etc.

The poets knew this long before it was confirmed by science. The proof is Rudyard Kipling's poem entitled *"The Two-Sided Man"*, which he wrote more than fifty years ago:

MUCH I owe to the Lands that grew -
More to the Lives that fed -
But most to Allah Who gave me two
Separate sides to my head.

Much I reflect on the Good and True
In the Faiths beneath the sun,
But most upon Allah Who gave me two
Sides to my head not one.

Wesley's following, Calvin's flock,
White or yellow or bronze,
Shaman, Ju-ju or Angekok,
Minister, Mukamuk, Bonze -

Here is a health, my brothers, to you,
However your prayers are said,
And praised be Allah Who gave me two
Separate sides to my head!

I would go without shirt or shoe,
Friend, tobacco or bread,
Sooner than lose for a minute the two
Separate sides of my head!

~ Rudyard Kipling

Dr. Betty Edwards, a world-renowned educator in the field of art and the author of the best-selling non-fiction books *"Drawing on the Right Side of the Brain"* and *"Drawing on the Artist Within"[1]*, does not agree with the notion that some people are just not artistic. "Drawing is not really very difficult...Seeing is the problem" she says. The secret to seeing is quieting down the bossy left brain so that the "subservient" right brain can do its artistic wonders. *"Drawing on the Right Side of the Brain"* is about turning off the left hemisphere by giving the brain a task that the left hemisphere cannot do – therefore allowing the right perceptual side of the brain to take over. Edwards has been able to prove this claim by running a five-day tutorial in which she has trained many ordinary people [without any artistic abilities] to draw.

Dr. Edwards has summarized the capabilities of the left and right brain in her books *"The New Drawing on the Right Side of the Brain"* and *"Drawing on the Artist Within"*. On page 44 of *"The New Drawing on*

---

[1] Dr. Betty Edwards, author of the best selling non-fiction books *"Drawing on the Right Side of the Brain"* and *"Drawing on the Artist Within"*

the Right Side of the Brain", these are surmised in the table shown below:

L –mode	R-mode
Verbal Using words to name, describe, define.	**Nonverbal** Using non-verbal cognition to process perceptions.
Analytic Figuring things out step-by-step and part-by-part.	**Synthetic** Putting things together to form wholes.
Symbolic Using a symbol to stand for something. For example, the drawn form stands for the eye, the sign + stands for the process of addition.	**Actual, Real** Relating to things as they are at the present moment.
Abstract Taking out a small bit of information and using it to represent the whole thing.	**Analogic** Seeing likenesses among things; understanding metaphoric relationships.
Temporal Keeping track of time, sequencing one thing after another: doing first things first, second things second, etc.	**Nontemporal** Without a sense of time.
Rational Drawing conclusions based on reason and facts.	**Nonrational** Not requiring a basis of reason or facts; willingness to suspend judgment.
Digital Using numbers as in counting.	**Spatial** Seeing where things are in relation to other things and how parts go together to form a whole.
Logical Drawing conclusions based on logic: one thing following another in logical order—for example, a mathematical theorem or a well-stated argument.	**Intuitive** Making leaps of insight, often based on incomplete patterns, hunches, feelings, or visual images.
Linear Thinking in terms of linked ideas, one thought directly following another, often leading to a convergent conclusion.	**Holistic** [meaning "wholistic"] Seeing whole things all at once; perceiving the overall patterns and structures, often leading to divergent conclusions.

The work activities involved in constructing the DW application are grouped into two categories: engineering-oriented and human-oriented activities. Comparing this with the activities grouped in Edwards' table above. Clearly, engineering-oriented activities fall under the left brain domain, while human-oriented activities are under right brain domain as below:

- Left-brain thinking: logical, automatic, rigorous thoughts characterizing most engineering jobs.
- Right-brain thinking: sentiments, feelings, inventiveness; characteristics of pioneers who see whole things all at once and people who think out of the box.

Software engineers know how to program, test and build DW. These are software engineering activities that been successfully performed for last few decades. To ensure success in the future, it is important to learn how to build bridges across the chasm between both hemispheres of our brain.

In general, Western society is facing a crisis of tectonic proportions. Professions such as engineering and programming, responsible for bringing prosperity to our homes, are now in serious danger as businesses move away from an economy and society built on the logical, linear, computer-like capability of the Information Age toward an economy and society built on the inventive, empathetic, big-picture capability of what is called the Conceptual Age - a new era, in which the mastery of abilities which we have often overlooked and undervalued marks the future divide between those who succeed and those who don't.

Management guru Peter Drucker gives the name "knowledge workers[2]" to:

> *"...people who get paid for putting to work what one learns in school rather than for their physical strength or manual skill."*

He writes:

> *"What distinguished members of this group and enabled them to reap society's greatest rewards was their "ability to acquire and to apply theoretical and analytic knowledge."*

Typically, these professionals include people like tax attorneys, radiologists, financial analysts, software engineers etc... They have taken the path to the professional success and personal fulfillment of the Information Age[3]. Unfortunately, the world has changed. A quote from D. Pink's book illustrates this nicely:

> *"A funny thing happened while we were pressing our noses to the grindstone: The world changed. The future no longer belongs to*

---

[2] Peter Drucker: *The Age of Discontinuity (1969)*

[3] Daniel H. Pink: *A Whole New Mind: Why Right-Brainers Will Rule the Future*

people who can reason with computer-like logic, speed, and precision. It belongs to a different kind of person with a different kind of mind. Today — amid the uncertainties of an economy that has gone from boom to bust to blah — there's a metaphor that explains what's going on. And it's right inside our heads."

~ Daniel H. Pink in "A Whole New Mind —
Why Right-Brainers Will Rule the Future"

Pink suggests nothing less than a paradigm shift. The world is changing before our eyes. We are moving from the Information Age into the Conceptual Age. The current Information Age in which access to information [knowledge], acquired mostly though education, was the most important economic engine is changing into the Conceptual Age, which is based on context, patterns, the big picture and emotions. As in the Industrial Revolution, during which many human jobs were replaced by machines, today too, any job that requires logical, analytical left-brain thinking is now being exported to the less expensive offshore workforce or replaced by a computer program. Many new professions are acquiring right-brain skills as it becomes harder and harder to keep the old job in the new era. The job market no longer belongs to knowledge workers [people who can reason with computer-like logic]. It belongs to a different kind of person with a different kind of mind. It belongs to creative individuals, those who can take the mountains of information and knowledge and create something with intrinsic value.

Left-brain skills are still absolutely necessary in the Conceptual Age, but they're just not sufficient any more. The key to success in the future will be the ability to use the whole mind. Today, our economy, based on a left-brain-skilled [white-collar] workforce, has gone from a splendid boom to a painful bust and will soon amount to nothing if we don't do something about it. It's actually a powerful metaphor which parallels what must be happening inside our brains too, if we want to continue to prosper in the Conceptual Age.

The Conceptual Age started long ago when humans began using symbols and gestures to refer to objects and concepts. The symbols and gestures are **not** the objects themselves; they merely represent the objects. Cognitive objects are cognitive tools that interact with the mind and they have an impact only on those who know how to use them. Even then, the impact itself is dependent on **who** uses them[4]. The world we perceive is not only raw optical input from our eyes, but rather

---

[4] Donald A. Norman: *Things That Make Us Smart* [page 47]

something constructed by our brain, using the optical input from our eyes and our lifelong experience and memory of the world around us. Thus, before producing cognitive objects, we must empathize with the user of the created objects.

Forrester Research predicted that 1 in 9 jobs in the US information technology industry would move overseas by 2010. And it's not just the high-tech work. India's chartered accountants are now preparing American tax returns, its lawyers are researching American lawsuits and its radiologists are interpreting CAT scans for US hospitals. Even if left-brain work doesn't move offshore, given ever increasing computing speed and capacity, it will be automated soon enough.

There is nothing we can do to stop the disappearance of white-collar jobs. Left-brain work, such as software engineering [or any engineering], accounting, legal research, and financial analysis are migrating across the oceans!

Whole-brain thinking isn't something new, but something old that has just gained importance. Civilizations have always tried to improve aesthetics. Athens - a democratic city-state which trained its citizens in the arts of both peace and war and gave the world the first literary masterpieces, the "Iliad" and the "Odyssey", the Byzantine city of Constantinople - whose astonishing beauty could not be destroyed by barbarians, Rome - once the epitome of civilization, and others, all stand as evidence that civilizations have always attempted to beautify their surroundings. They designed new architectural styles and made their buildings more pleasing. Right-brained thinking isn't a new way of thinking. It has always been inherent in human nature.

Upon the first realization of this trend, there was anger – how dare they to export our jobs to foreign countries? Then came the rationalization: our national companies are competing in the world marketplace, and in order to stay competitive, they must seek a way to lower their operational costs. The question is: who do we save, our outdated jobs or our modern companies? Work in any occupation, no matter how diligently we follow it, can lead to staleness if we lose our horizon and the excitement of those horizons. If we don't embrace the new reality, we will soon face the grief of jobless, unwanted software engineers.

As increasing numbers of high-tech left-brained tasks gradually move overseas to countries able and willing to perform the same work for a fraction of Western salaries, we shouldn't focus on preserving those jobs, as they have become commodities. Instead, we should focus on new opportunities arising in their place - jobs that require abstract, holistic, intuitive skill sets blending aesthetic beauty into harmonious systemic thinking and in process creating a culture of invention that cannot simply be automated by computer programs or outsourced to the lowest bidder.

It is said that "every cloud has a silver lining", and we will be left with exciting new job opportunities that require whole-brain thinking!

Psychologist Dr. Mihaly Csikszentmihalyi's famous investigations of "optimal experience" have revealed that what makes an experience genuinely satisfying is a state of consciousness called *flow*. We are in this state of "flow" when our whole brain is engaged in thinking, when people typically experience deep enjoyment, creativity, and a total involvement with life. In the new edition of his groundbreaking classic work, Csikszentmihalyi demonstrates ways in which this positive state can be controlled and not just left to chance.. His book: *Flow: The Psychology of Optimal Experience"* [5] teaches how, by ordering the information that enters our consciousness, we can discover true happiness and greatly improve the quality of our lives. The "flow" research is deeply rooted in Ancient wisdom starting with Aristotle's view on the good life. Aristotle's model of the good life is "Eudaimonia", "the state of being well and doing well," often translated as "happiness". According to the Aristotelian concept of happiness, it is not all the other things we seek - riches, fame, power, etc - that make us happy, but we pursue them only because we believe they will make us happy.

Two conditions are required for the "flow": the challenge and the skills to meet the challenge. We now have the challenge and it's up to us to supply the necessary skills to achieve something difficult and worthwhile. In the process of building our "field of dreams", we are, at the same time, creating opportunities for our national companies to open up new, exciting jobs which are now going unfilled due to the lack of qualified people, and simultaneously fulfilling their purpose of existence – to create national wealth at home.

---

[5] You must have heard about how a musician loses herself in her music, how a painter becomes one with the process of painting. In work, sport, conversation or hobby, you have experienced, yourself, the suspension of time, the freedom of complete absorption in activity. This is "flow," an experience that is at once demanding and rewarding - an experience that Mihaly Csikszentmihalyi demonstrates is one of the most enjoyable and valuable experiences a person can have. The exhaustive case studies, controlled experiments and innumerable references to historical figures, philosophers and scientists through the ages prove Csikszentmihalyi's point that flow is a singularly productive and desirable state. But the implications for its application to society are what make the book revolutionary.

We all need purpose and meaning in life, as Viktor Frankl said in his book *"Man's Search for Meaning"* [paraphrasing]: only those that had something else in their lives that had yet to be done survived the most terrible conditions that one human being can put another through - concentration camps. He must know as he survived imprisonment in Auschwitz and other concentration camps for five years, struggling during this time to find reasons to live. Frankl believed that one's deepest desire is not pleasure, as Freud maintained, but to search for meaning and purpose. Meaning in life allows us to have a drive and ambition to achieve the highest goals. We need to spend quiet time in meditative thought and to develop our minds and exercise our imagination.

As the old saying goes, "an idle mind is a workshop for the devil". The most prolific inventor of the last century Nikola Tesla [Serbian: Никола Тесла; 10 July 1856 – 7 January 1943], whose patents and theoretical and practical work on the modern alternating current [AC] with his polyphase AC motor gave the world a complete solution for the production and distribution electric power known as, the man who "powered" the 21st Century, Tesla said:

> *"The mind is sharper and keener in seclusion and uninterrupted solitude. Originality thrives in seclusion free of outside influences beating upon us to cripple the creative mind. Be alone—that is the secret of invention: be alone, that is when ideas are born."* •

Motivation is the driving force which causes us to achieve goals. The term motivation is generally used for humans but theoretically, it can also be used to describe the causes of animal behavior. In motivating us to work for its goals, society is assisted by the powerful forces of nature: our biological needs and our genetic conditioning. For example, cravings based on our genetic programming, from sexuality to aggression, have been exploited as the basis for control by those in power. Survival instinct is exploited in oppressed societies for purpose of social control. When society no longer responds to pain, the use of pleasure is employed. .

The two motivational models proposed, by Freud and by Frankl are exemplified below and left alone to stand on their own merits:

Csikszentmihalyi in his book *"Flow: The Psychology of Optimal Experience"* describes an example of physical motivational "technology" invented by the Ottoman Empire:

> *"To lure recruits into the Turkish armed forces, the sultans of the sixteenth century promised conscript the rewards of raping women in the conquered territories."*
>
> [*Flow: The Psychology of Optimal Experience*, Mihaly Csikszentmihalyi, 1990, page 17]

Contrast this kind of motivation based on physical desires with motivation based on psychological needs. Some examples are, the monumental architecture of artists of the Baroque era that can still be seen today in Rome or Vienna that transcend time and place. Baroque music, best represented by composer Johann Sebastian Bach, whose music [6] still reverberates in music halls around the globe. The mathematical precision and harmony of the artistic creation of the Baroque era is striking. How could these artists have created such stunning and timeless beauty? They must have been totally absorbed by their creation. Since they did not work for money as they were mostly poor, what motivated them? Art critics maintain that the art of the Baroque era was meant to honor **God**!

Until recently, Software engineers have been expected to produce facts coolly [without emotions] and objectively, uncolored by their personal feelings. This approach is no longer enough, as this kind of thinking does not produce cognitive objects that would stimulate cognitive thinking in business. Cognitive objects [i.e. reports and graphs] can be viewed of as objects being stored in a place of temporary memory external to the mind. By using these cognitive objects which are displayed in the external memory of the human mind, the biological limitation of the mind is being expanded. People on the receiving side of the object must be able to interpret the "script language" used in the objects. Consequently, the ability to see the world "with someone else's eyes" [understanding other human beings - empathy] is a crucial criterion for engineers, who without it would not be able to provide the appropriate cognitive objects that would stimulate the desired insights to the viewer. Humanistic skills must be mustered before engineers can craft a story from the data locked in the DW that will resonate in symphonic unity with the business user's mind.

Twenty-three centuries ago, Aristotle wrote that more than anything else, men and women seek happiness. There is now a strong body of

---

[6] Johann Sebastian Bach, George Frideric Handel, Alessandro Scarlatti, Antonio Vivaldi, Jean-Baptiste Lully, Arcangelo Corelli, Claudio Monteverdi, Jean-Philippe Rameau and Henry Purcell.

evidence, based on the last decade of research in the fields of psychology and neurology, applied to business that offers a formula equating happiness with success. In other words, the success rate is directly proportional to the level of happiness! Our brains work better when they are "happy"[7].

Contrary to popular perception, our natural tendency is not to pursue happiness. Our default response is inertia - we continue doing what we have been doing, expecting a different outcome. In reality, we even know that won't make us happy [remember the last time you wasted a whole day in front of the television even though you had plans that would have given you more happiness and meaning?]. Things that make us happy, like laughing with someone we love, or discovering a perfect piece of poetry, or searching for what is truly beautiful, are the same things that make us more creative and more productive! Interesting... would that not be a strong incentive to make "happiness" a part of the work ethic?

Happiness is not something that money can buy. It is not dependent on external situations, but rather on our mental perception of the situation. Drucker illustrates this in "The Daily Drucker: 366 Days of Insight and Motivation for Getting the Right Things Done"[8], in a story about stonecutters; when three of them were asked what they thought they were doing, one replied: "I am making a living"; the second replied: "I am doing the best job of stonecutting in the entire country" and the third said: "I am building a cathedral". How you perceive what you do makes a big difference in your happiness level.
Today, religion is not a very good motivator, for whatever reason. But it is up to each of us to find our own motivation, as success in business and in life requires 100% motivation and 100% devotion. It is actually very interesting that those are the same requirements that could snap us out of lethargy and into happiness!

To conclude, the left brain thinking jobs are gone! What remains are very exciting higher-level jobs that require whole-brain thinking. However, it is a fatal mistake to assume that they are ours by default. Especially other societies have a long tradition in this way of thinking. To name a few examples: the yogi discipline in India[9], the Zen varieties of Buddhism[10],

---

[7] DR. Tal Ben-Shahar: *Happier: Learn the Secrets to Daily Joy and Lasting Fulfillment*
[8] Dr. Peter Drucker: *The Daily Drucker: 366 Days of Insight and Motivation for Getting the Right Things Done*
[9] Yoga is the ability to direct the mind exclusively toward an object and sustain that direction without any destruction.

the Taoist approach to life developed in China[11], etc. Instead of being in competition with these societies, we should embrace these differences and direct our efforts to cooperating with them hoping, that by virtue of synergy, something good, something much better, will evolve.

Fig 4.1 Yin Yang Fish

---

[10] Zazen is a Japanese term consisting of two characters: *za*, "to sit (cross-legged)," and *Zen* meaning is at once concentration, dynamic stillness, and contemplation. The means toward the realization of one's original nature as well as the realization itself, Zazen is both something one does - sitting cross-legged, with proper posture and correct breathing - and something one essentially is.

[11] The Chinese characters for Tai Chi Chuan can be translated as the 'Supreme Ultimate Force'. Tai Chi, as it is practiced in the west today, can perhaps best be thought of as a moving form of yoga and meditation combined

Quality Assurance Story

"It is not enough that the top management commit itself for life to quality and productivity. They must know what it is that they are committed to---that is, what they must do. Those obligations cannot be delegated. Support is not enough; action is required."

~ Dr. W. Edwards Deming

One spring day, after the winter winds had died down, after long cloudy night skies finally gave way to the stars and the moon, the most extraordinary thing occurred.

The king woke up from the in his dream in which he heard the voice telling him the most amazing story about this great land that had elephants. Next morning, he summoned his three wise advisors. He told them a story that he had heard in his dreams, about a great land that had elephants. In spite of the fact that he had not seen an elephant and did not know what kind of animal an elephant is, he was sure that this was just what he needed to save his country from its poverty.

The king's advisors had not seen an elephant before either, but they were sure they could get him some. Since they did not know what an elephant is, they travelled to this distant land to see it. They arrived in the great land in the middle of the night. They were tired from their long journey, but they insisted on seeing an elephant as soon as possible, so that they could go home and give their king one or even more of these elephants. They were taken to the Elephant House in the middle of the dark, moonless night. The first advisor entered the dark room and began to feel around for an elephant. He touched the elephant's leg. He immediately declared: it's a tree! The second advisor touched its tail and declared: it's a snake! The third advisor bumped into the elephant's side and it felt exactly like a wall. He said to himself: it must be a wall!

The three advisors went home and declared themselves to be the "Elephant Masters". Soon, they started roaming the land, giving expensive advice on how a person could find an elephant on their own.

The king, disappointed that he did not get his elephant, on which he had placed great hopes for his land, ordered the three advisors to be locked in the "Elephant Institute" every day from dawn till dusk, until they gave him and the people of his land an elephant, or until the age of 65.

Quality Assurance is not a thing that one can somehow be gotten. Nether is it a destination that one arrives at. Quality Assurance is the commitment to making a long journey! It is the road less traveled - the road that twists and turns and no two directions are ever the same. There is no absolute definition of Quality Assurance, as the processes can always be improved and taken to the next level.

To pursue Quality Assurance is to reject the complacency of thinking that we have finally "gotten it perfect". Because there is no final end, it's a moving target – a journey, as our joy is found not in finishing an activity, but in doing it.

Ithaca is the island in Greece that Odysseus had so much trouble returning to, in Homer's Odyssey. Greek poet Constantine Kavafis wrote a poem called Ithaca in 1911 explaining it. Here is a part of it

> Keep Ithaca always in your mind..
> Arriving there is what you're destined for.
> But don't hurry the journey at all.
> Better if it lasts for years,
> so you're old by the time you reach the island,
> wealthy with all you've gained on the way,
> not expecting Ithaca to make you rich.
>
> Ithaca gave you the marvelous journey.
> Without her you wouldn't have set out.
> She has nothing left to give you now.
> And if you find her poor, Ithaca won't have fooled you.
> Wise as you will have become, so full of experience,
> you'll have understood by then what these Ithacas mean...

Defining quality is hard because everybody seems to have their own preconceived idea of what quality means, but for the most part, they disagree amongst each other. It is like in the above parable, frequently told in management, about three wise men that came across an elephant and could not agree on what an elephant is. Quality, too, may mean different things to different people. Consequently, when two people are discussing quality, often they are talking about different things. A customer-oriented user may be looking for ease of use or quick transaction time. A more technical "knowledge user" may be looking for a rich set of functions. A project manager may be looking for all of the above and for delivering it on time and within the budget.

The quality of DW application is an elusive aspect of it, not because it is hard to achieve [once we agree what it is], but because it is difficult to describe.

In the context of this book, we do not see quality according to the classic definition by Gause and Weinberg: "*A problem can be defined as the difference between things as perceived and things desired.*" Instead, we propose the notion that quality is not an attribute or a feature of a product but rather a *relationship between that product and a stakeholder.* More specifically, the relationship between the software quality and the organization that produces the products is explored. This is a multi-faceted relationship, dependent on many factors - business strategy, business culture, available talents, processes that produce the products, etc. We are purposely using a slightly more inclusive view of quality, as quality, especially in DW application, is more elusive and highly unpredictable.

The conceptual framework for software quality was built on the pioneering work of Dr. W. E. Deming. It was his leadership in statistical quality control that propelled Japanese industries [automobile, shipbuilding, electronics, and others] to world prominence back in the 1950s and 1960s. More recently, the performance of the Japanese software industry is showing that these concepts of statistical process control are as applicable to software development as they are to producing consumer goods like cameras or television sets. The basic principle of statistical control is measurement. It is said that "what gets measured gets improved", or as Lord Kelvin said one century ago:

> "*When you can measure what you are speaking about, and express it in numbers, you know something about it; but when you cannot measure it, when you cannot express it in numbers, your knowledge is of a meagre and unsatisfactory kind; it may be the beginning of knowledge, but you have scarcely in your thoughts advanced in the stage of science.*"

It is the function of a QA Team to plan a systematic pattern of actions that will provide the confidence that all the stakeholders' expectations are met and at the same time, ensuring that DW application project standards and procedure are maintained throughout all the phases of the project and ensuring the objective of QA in preventing defects from being built into products! Consequently, quality is built into the products at all stages of the SDLC, but when it comes to testing, it is too late! Nevertheless, the goal of QA is not QA activities and processes, but the awareness of everybody who works around these QA processes that the goal of QA is to deliver a tool to business users which will enable them to better steer the ship through the stormy sea in front of us.

From the quality assurance point of view, it's not how well one can measure the quality of the product, even though this is an important activity but they will not achieve the desired product quality. Measure of product quality is just feedback [a learning opportunity] of how effective the software quality program that produced that product is! Quality is affected by many; however, only those that are actually producing the product build the quality into it! QA actions and methodology must be focused on integrating the efforts of the many and the few, bringing them to achieve the objective of quality together.

The software quality program and overall approach in achieving the desired quality are laid in and are an integral part of the Test Strategy document. The activities necessary to achieve the quality objectives are described in it:

- Establish clear, objective and verifiable requirements for the product quality.
- Implement and enforce methodologies, processes and procedures to support the quality objective.
- Implement validation and verification methodology, processes and procedures to evaluate the quality objective throughout all the phases of the SDLC.

This book concentrates on differences in QA processes between the DW Application and classic software application projects. Software Quality methodology, such as the CMMI [Capability Maturity Model Integrated] is still required but not sufficient for DW application projects. The fundamental notion in CMMI is the management premise that:

> *"The quality of the system or product is highly influenced by the quality of processes used to develop and maintain it".*

CMMI provides a comprehensive description in different stages of maturity and the conditions that determine where one is and where one can hope to be in order to grow, thus turning the corner from chaotic software development to a controlled management process. While it may seem trivial to define the current state of software development processes in an organization, it is not. One way to assess the software capability of an organization is to watch what it does in a crisis. That is when the good practices are most critical and when guidance is required.

If the processes in use by the software development project are not defined and organized well, then the quality of the software product is nether predictable nor repeatable. The dependence of software quality and processes that are producing the software products is the basic assumption of the Software Engineering Institute [SEI] in their CMMI model for software development. Five levels of maturity are described by the CMMI:

Level	Characteristics	Key Problem Area
Initial	Processes are ad hoc and chaotic	Project management Project Planning Configuration management Software quality assurance
Managed – Repeatable	Process is managed in accordance with agreed metrics	Training Technical practices: - reviews, testing, etc. Process practises: - standards, process groups, etc
Defined	Processes are well characterized and understood, and are described in standards, procedures, tools, and methods	Process measurement Process analyses Quantitative quality plans

Quantitatively managed	Quantitative objectives for quality and process performance are established and used as criteria in managing processes	Problem causal analyses Problem prevention
Optimizing	Process management includes deliberate process optimization/improvement	Process automation

Low maturity software organizations spend most of their time and energy fighting fires in trying to contain a high volume of changes or recovering from late-discovered defects found in production. The same defect is rediscovered many times and many solutions are reinvented as a management process is non-existent. Any organization where the heroic efforts of a lone cowboy, rather than an organised QA process of defect prevention, are more popular with the management is also risky place to work. When a crisis hits, their solution is the technical wizard; unfortunately there is often no way to recover. As Dr. F.P. Brook says, "There is no silver bullet" [1]. The solution is systematic project

---

[1] Dr. F.P. Brooks *"The Mythical Man-Month: Essays on Software Engineering"* –
The classic book on the human elements of software engineering
Some of Brooks' insights and generalizations are:
The Mythical Man-Month: Assigning more programmers to a project running behind schedule may make it even more late.
The Second-System Effect: The second system an engineer designs is the most bloated system she will EVER design.
Conceptual Integrity: To retain conceptual integrity and thereby user-friendliness, a system must have a single architect (or a small system architecture team), completely separate from the implementation team.
The Manual: The chief architect should produce detailed written specifications for the system in the form of a manual, which leaves no ambiguities about any part of the system and completely specifies the external specifications of the system i.e. what the user sees.
Pilot Plant: When designing a new kind of system, a team should factor in the fact that they will have to throw away the first system that is built since this first system will teach them how to build the system. The system will then be completely redesigned using the newly acquired insights during building of the first system. This second system will be smarter and should be the one delivered to the customer.
Formal Documents: Every project manager must create a roadmap in the form of formal documents which specifies milestones precisely and things like who is going to do what and when and at what cost.

management, where work is estimated, planned and managed and where quality assurance is an independent team, charged with the responsibility of assuring that all the processes are properly performed.

For an excellent discussion of the CMMI process, we refer interested readers to the works [Ref. 41] of Watts S. Humphrey [also known as the "father of software quality"], a fellow and a research scientist at the Software Process Program of the SEI at Carnegie Mellon University in Pittsburgh. He wrote the first version of the SEI's CMMI for software in 1987. Humphrey also developed the Personal Software Process [PSP] and the Team Software Process [TSP]. His work was based on Dr. W. Edward Deming's principles of statistical process quality control, summarized in Deming's 14 management principles [Ref. 42] and Humphrey's strong practical background at IBM, where he worked from 1959 to 1986 as a director of programming quality and processes, including on IBM's most successful OS/360 project. We are associated with neither SEI nor Mr. Humphrey, but the above credentials are strong reason for recommending his work.

A fundamental problem, however, with the CMMI or any other QA software engineering methodology in development of DW application or the most important part of it, mainly Business Intelligence, is that BI is more in the realm of art than it is in engineering and the more your try to control it, the more it disappears!

Before continuing with the descriptions of various QA processes, it is important to stress that everyone on the project communicate using the

---

Communication: In order to avoid disaster, all the teams working on a project, such as the architecture and implementation teams, should stay in contact with each other in as many ways as possible and not guess or assume anything about the other. Ask whenever there's a doubt. NEVER assume anything.

Code Freeze and System Versioning: No customer ever fully knows what she wants from the system she wants you to build. As the system begins to come to life, and the customer interacts with it, she understands more and more what she really wants from the system and consequently asks for changes. These changes should of course be accommodated but only up to a certain date, after which the code is frozen. All requests for more changes will have to wait until the NEXT version of the system. If you keep making changes to the system endlessly, it may NEVER get finished.

Specialized Tools: Every team should have a designated tool maker who makes tools for the entire team, instead of all individuals developing and using their private tools that no one else understands.

same terminology; otherwise a project may end up like the ancient Tower of Babylon:

> *Up until that point in the Bible, the whole world had one language - one common speech for all people. However, the people of the earth became skilled in construction and decided to build a tower that would reach to heaven. God came to see the tower they were building. He perceived their intentions, and in His infinite wisdom, He knew this "stairway to heaven" would only lead the people away from God. He noted the powerful force within their unity of purpose. As a result, God confused their language, causing them to speak many different languages so they would not understand each other. By doing this, God thwarted their plans. That is why the Tower of Babylon collapsed.*

Fig. 5.1 Tower of Babylon

Software Testing vs. Software QA

Let's start by clarifying the differences and distinctions between software testing and software QA. The two terms have often mistakenly been used synonymously, but they are two distinct processes, which, however, are complementary and if properly employed, synergetic. They have different goals and require a different skill set. QA is a management/leadership activity responsible for setting up and enforcing processes and methodologies in order to prevent defects. Defect prevention is a powerful technique for improving software process. The general idea is to track every defect and periodically perform causal analyses. Depending on the project and the organization, there are number of standards [CMMI, ISO, IEEE, etc.] that can be deployed in the software development process of DW application.

Testing, on the other hand, is a technical activity responsible for finding defects - all of them and sooner rather than later. If the QA describes "what needs to be done", testing describes "how it is going to be done". Testing is the subject of another chapter in this book.

Software failure is the inadequacy of the system to execute a required function within specified time. It is manifested by incorrect output, abnormal program termination or system crash. Software fault is caused by incorrect or missing code. A bug is a software fault too. It was first used to describe a problem when the moth got wedged in the computer relay causing the program to halt. A bug is corrupting euphemism. A bag should not be called "a bug", as it metaphorically blames the cause of the fault to some evil forces beyond the programmer control. It is much more honest to call it an error or a defect as it is created by the programmer's action. This slightly changes the perspective on the program that does not work and places the responsibility where it belongs.

Validation vs. Verification

Other terms that are also often mixed up are "Validation" vs. "Verification" [V & V]. Definitions as used in this book are presented here:

The purpose of validation and verification [V & V] is to establish with confidence that the Data Warehouse is fit for its intended use. However, this does not guarantee that the warehouse application is completely free of defects.

The Validation and Verification testing processes must be integral part of every phase in the entire Software Development Life Cycle [SDLC]. They must be applied at each stage in the SDLC, with principal objectives including:

- The continuous assessment of whether or not the DW Application is useful and useable in an operational situation by business users
- The prevention and discovery of defects in the DW Application

Validation

First we do the Validation to answer the question: "Are we building the right product?" In other words, does the software do what the user actually requires?

Business users of the Data Warehouse Application must be provided with all the objects, such as, use cases, data flow diagrams, VISIO diagrams, and mock-up screens and reports based on "near-production data", etc. This enables them to fully understand the solution and ultimately validate that the EDW under development complies with their requirements, performs functions for which it is intended, and meets the organization's goals and user needs. A typical reaction when the user sees the application in production for the first time is:

> "Yes, this is nice, but now that I see how it works, wouldn't it be better if..."

Avoiding the "yes, but" syndrome is the reason for having the user involved in the validation process. The root of the "Yes, but" syndrome is the inherent intangibility of the intellectual process. The "Yes, but" syndrome can be avoided by providing users with all the intermediate objects and keeping them involved in the validation and decision-making process. It is important to keep in mind that the user being "busy" is not an acceptable excuse since "it's always cheaper to build the application right the first time, than to do it all over again".

Verification

Only when we know what we are going to build do we proceed with the verification. "Are we building the product right?" is a question that we must answer during the verification process, or translated into software parlance "does the software conform to its validated specification?"
Verification is performed during development on all key deliverables, like walkthroughs, reviews, inspections and testing. The goal of verification is to demonstrate the consistency, completeness and correctness of the DW application at each stage and between each stage of the software development life cycle.

Verification ensures the adherence to validated user requirements, the goal being determining if the system is consistent with validated requirements and if it adheres to standards and performs the selected functions correctly.

Both effective and efficient software processes are critical to successful EDW implementation, but their merit can only be determined in the context of the business needs of the particular organization.

Classical Software Development Life Cycle [SDLC] revolves around the "Waterfall" or "V" model.

V & V are depicted and marked as appropriate to that model in the picture below.

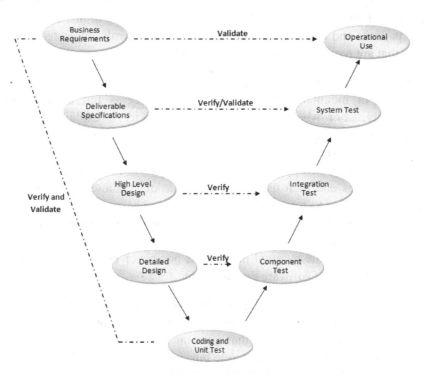

Fig. 5.2 Waterfall SDLC

What's different?

> *"Everything should be as simple as possible, but not simpler"*
> ~ Albert Einstein

Complexity of data is being recognized more and more as the key characteristic of the world we live in. The size of data compounds the problem of complexity by masking the hidden patterns upon which the complexity is built. Sometimes, data complexity is built upon the simple pattern of data redundancy. Seldom are we interested in uncovering all the patterns and relationships of data, but we are mostly interested in the few that are required to answer questions posed by businesses.

The central task of a BI data analyst is to show, by finding hidden patterns in mountains of data, that complexity is but a mask for simplicity. For when the wonder of complexity is explained, using aesthetics of art and mathematics by discovering concealed patterns, a new wonder arises of how complexity was woven out of simplicity, thereby creating the opportunity for data analysts to first find and analyze these patterns and then use them, by virtue of syntheses, in crafting a convincing story to form the new complexity. These uncovered patterns and relationships of data are presented to users as cognitive objects [graphs and reports] and can be thought of as an "interface" between the user's mind and the external data. This has never been done before in the short history of software development.

The development of DW application evolves at the junction of two worlds – the IT world and the business world. DW application is not a project but rather a program that goes on with no endpoint or finished deliverable. Yes, there are concrete business requirements, but to plan it as typical software deliverable that is "finished" at any point in time should not be an expectation. DW application is a project that should live as long as the business needs information. It does not follow any traditional software life cycle of birth, life and death.

The DW application continues to evolve over time, dissimilar to a software development project that is complete when the project is deployed. Software application testing methods focus mostly on testing transaction-oriented systems. They involve code testing, while DW application testing is different in many respects. Yes, classic software testing is an important part of DW application testing, but it is not the only part.

The DW application development process is highly iterative and without a finite end, therefore requiring continuous testing. The scope of testing goes beyond just software testing. It starts with testing the IT alignment with business goals and ends by testing delivered business value.

Business requirements for the DW application projects are usually fuzzy, indefinite and incomplete, less stable and more prone to change. The potential for additional requirements that may affect the design in the early development cycle is very high. This is because when the user is presented with the solution, he will recognize the potential of the technology. That is why it is very important to have the business user involved in the project from the beginning.

While software application testing has the goal of ensuring the functional correctness of the code, DW application testing focuses on the business value of the information derived by the delivered data. Information quality goes beyond data quality. Special consideration should be given to data quality, and perhaps the most important aspect of data quality – data appropriateness.

We often measure and collect information about those things that are easy to count. This may give us a false sense of security, but by ignoring those intangible [hard to measure] factors we may be "missing the forest for the tree". For instance, it is much easier to measure the quality of a product than its impact on the quality of life. We measure what we can and ignore the rest. By ignoring what is left, we may be ignoring a critical part of a typical company that is based on intangible assets. Consequently, what gets measured gets improved, because we value what we can measure. We may be doing a superb job on something of little consequence and leaving out the most critical part. This is why we say that before we start being efficient, we first must make sure we are effective!

The foundation of software application testing is found in business requirements and system designs. The requirements for building and testing DW applications are much more elusive. The objective of software application testing is to verify the correctness of the program code and that it does not fail [and if it does fail, it does it gracefully]. This must be done before the application is deployed in production. DW application testing is directed at data and information testing and it goes on [should go on] long after the DW is deployed.

Implementing a DW application initiative is not a project, but a long-term commitment to implementing continuously improving business intelligence practices. Consequently, a DW application development process is highly iterative and without a finite end, therefore requiring continuous testing.

Data warehousing applications use a distinctly different SDLC model, illustrated in the diagram below. DW application development is a new paradigm and it requires a new SDLC. We are proposing the model that we have been using successfully – Perpetual [never ending] Iterative SDLC. The name is fair to our businesses; they should also be made aware that these [DW] projects never end! At the heart of this process is validation, as can be seen in the picture below.

Validation at every step of the cycle!

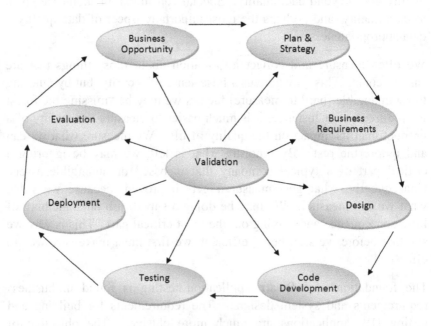

Fig. 6.3 DW SDLC Model

The differences discussed above dictate changes in the:

1. Structure of the QA and development teams
2. Software Development Life Cycle and
3. Nature of testing.

How many times have we heard business people telling us:

> *"I don't need all those reports—just give me the numbers that I need, to find out what's going on."*

It is the IT people that determine what those numbers that the business user needs are. Even though these are business professionals who have gone to different universities than the IT profesionals don't know anything about their day-to-day work, they still want to provide them with reports that they need in order to do their work!

Programming is intellectually demanding and all-absorbing work that dominates over all other considerations, including alien thought processes concerning users. Programmes want to do a good job, as everybody else does. So what do programmers do without full requirements? If for example, programmers are asked to create an accounting system report, they go and read some accounting books and magazines, find some fancy graphs and reports, and then improve on them - "improve" on something they do not know anything about[!] by adding bells and whistles to dazzle [intimidate] the business users. Correct reports have been delivered [except that no one knows if they are usable] and the project is completed and closed. But without knowing if the project has delivered the ultimate purpose of business intelligence - the story (cognitive objects that stimulate cognitive thinking) - that conclusion cannot be made.

The quote below is very appropriate for the state of DW application projects today:

> *High-tech companies—in an effort to improve their products—are merely adding complicating and unwanted features to them. Because the broken process cannot solve the problem of bad products, but can only add new functions, that is what vendors do...The high-tech industry has inadvertently put programmers and engineers in charge, so their hard-to-use engineering culture dominates...Programmers aren't evil. They work hard to make their software easy to use. Unfortunately, their frame of reference is themselves, so they only make it easy to use for other software engineers, not for normal human beings...While we let our products*

frustrate, cost, confuse, irritate, and kill us, we are not taking advantage of the real promise of software-based products: to be the most human and powerful and pleasurable creations ever imagined....All it requires is the judicious partnering of interaction design with programming.

[*The Inmates Are Running the Asylum*, Alan Cooper, 1999, SAMS Publishing: Indianapolis, Indiana, page 8, 15, and 17]

If you ask, as we have, any QA professional who has delivered DW application to business:

"How do you know the reports you have delivered are useful? You know the reports are correct, but are they useful to your users?"

Typical responses would be: "I know because our users are not reporting any defects!" In other words, reports are correct but [probably] useless! This is not the exception, it's the norm in the BI industry.

The traditional software vendors that are providing correct and effective data warehouse solutions are not able to move into the realm of visual analytics, which is where the information can be analysed, explored and used to predict the future. Success in the BI industry has come from a few non-traditional vendors like Tableau, TIBCO Spotfire [spin-offs of universities and research] and companies like SAS with a long history of work in statistics.

Is it any wonder that the "Yes, but..." syndrome keeps rearing its ugly head? That is why some authors are suggesting that the first delivery of BI reports to business users is a throwaway and a good place to start collecting real business requirements. We disagree, as the next iteration will suffer from the same disease.

Much more than that has to change, as this is not where the problem is, because the subsequent iteration will be a throwaway as well. We know how to collect, clean, transform and store data, but we don't know how to deliver information that can be analysed and used to predict the future! This is where our QA methodology differs from all others.

Quality Assurance Strategy for DW Application

"If you can't describe what you are doing as a process, you don't know what you're doing."
~ Dr. W. Edwards Deming

We begin with the end goal in mind, but the goal without a plan to achieve it, is just a dream. That is why we create a plan which prescribes **"what needs to be done"** in order to achieve the end goal of successfully deploying the DW Application. The plan is typically laid out in a document called the "Test Strategy" [or "Master Test Plan"]. The "Strategy" is the term we prefer to use in this book as the term metaphorically relates to the word "war", which conceptually relates to the brain the importance of this document.

Testing Strategy for DW application, first of all, must be aligned to business strategy. The very first priority of the strategy is mapping the top business questions into a problem definition and to achievable testing strategy objectives. But before starting to resolve the problem, the problem has to be defined first. More specifically, there must be process to ascertain what the key business question is. Only then we can start mapping the business problem to data visualization solutions.

Identifying the business problem is a team effort, between business users [domain expert], Data Analyst and Business Analyst. QA must be privy to this effort in order to be able to map the successful testing strategy. As the requirements for the success of the DW Application are vague and abstract, very difficult to visualize even by business user, some kind of prototype must be employed for continuous verification of the 'poof-of-concept'. Success criteria must include terms like, Return-on-Investment [ROI], profit and proposed matrix to track those.

Key Performance Indicators [KPIs] should be employed and agreed upon by all the stakeholders. KPIs are most of the times confused with performance target values. Use of KPIs in Business Intelligence is to assess the present state of the business and to prescribe a course of action. The KPIs are dependent on the nature of the organization and the organization's goals and mission. KPI are tactical tools used for "*mid-course evaluation & correction*". They help an organization to measure progress towards their business goals, especially toward difficult to quantify knowledge-based goals. KPIs can be expressed in many ways like, cost, quality, time, units of service, etc. These can be expressed in time, volume, money. KPIs can be measured in, absolute numbers, ratios, or a process improvement units.

Strategy defines all the processes, and which processes, must to be performed at which point of the project SDLC, in which environment and by whom, in order for DW Application to deliver expected business value. The necessary ingredient of the Testing Strategy is to define what success means, is success criteria [how do we know we have succeeded]. The definition for success of the organization must be directly related to success of the Testing Strategy.

Elements of the Strategy document are lists all the resources and tools, type of tests, design and testing documentation reviews, project assumptions related to QA, risk identification and risk mitigation, timelines and processes required to achieve the end goal of delivering quality product, on schedule and within budget and to the expectation of the business user.

Testing is the most critical element of the strategy. There are several types of test, but from our experience four have been the most valuable in DW application testing: unit tests, functional tests, integration tests and regression tests. These are explained further in the later part of this chapter. Automation testing, a special category of regression test is described and demonstrated in the last chapter of the book.

Risk and risk mitigation are also part of the strategy document. At the early sage of a project all the assumptions must be validated. For an example, one of the project assumptions may be that the company has all the data it needs to populate DW. If, however, the firm wants to build the data warehouse, but it is still using double-entry accounting system, invented 500 years ago, it is hard to believe that this company may have data it needs for the DW application. Double-entry accounting system was invented by Fra Lucia Pacioli 500 years ago, in Venice Italy and documented in his book *"Everything about Arithmetic, Geometry and Proportions"*. Double-entry accounting, based on tangible assets, was good then, but it is not sufficient any more today, where the market value of many modern companies [i.e. Google, Microsoft, etc.] is mostly based on intangible assets. It is unlikely that the company using double-entry accounting system would have data required for BI.

The greatest risk in developing DW project is delivering DW application that the business users can not use. This book presents the approach, to guards against that risk.

The same QA methodology that is applied to any software systems can also be applied to DW applications quality assurance. The QA testing cycles are the same; we still must execute all the regular testing, including, unit, integration, systems, performance [stress] and User Acceptance Testing [UAT]. However this is where the similarities between software application and DW Application testing strategy end.

Validate, Validate, Validate!

As already mentioned, first we have to know what it is we are building, before we start building it. Early and continuous involvement in testing provides the opportunity for early error detection and prevents migration of errors from requirement and specification to design and thence from design to being built into code. The earlier in SDLC the errors are uncovered the easier and less costly they are to fix.

Ours is the era of predominantly visual perception. Messages are received through the eyes first. The eyesight needs the shortest possible time, in comparison with other channels, to make impression on the brain. Vital quantitative business information is communicated to the brain in form of tables and graphs. Ggraphs are being used today, in every project and on every possible occasion, to convey messages and stores effectively, sometimes they are simple line or pie charts, sometimes to present complicated correlation within the data.

One of the functions of QA is to verify the usability of the delivered products. User requirements, in DW applications, are more elusive than ever before, but QA is also responsibility to validate these requirements and convey them unambiguously to the development teams. The QA methodology, we are about to lay down before you, is our original approach and it has been proven in our practice with many of our clients. We know that as we are often able to see that a client has produced tangible success as a result of our services. We have plenty of empirical evidence of the beneficial outcomes of this approach, but we are not allowed to share them publicly. In any case, we hope to be able to convince our readers, by the end of this cheaper, that this QA methodology will, shorten your development life cycle and diminish the possibility of errors. It is, at least, quicker and less expensive to do it right the first time, instead of repeating the entire effort all over again. This approach will, we are convinced, become ubiquitous in future. Absolutely essential and at the heart to this approach are two processes:

- Cognitive Data Visualization and
- Prototyping.

"Just one more thing" [Jobs] Have we said that this methodology will not work unless, business assume the responsibility of assuring the usability of the delivered DW application? As the domain expert knowledge is within business community, the responsibility of assuring the usability of the delivered BI application must clearly be placed on the business user's shoulders.

Data Visualization

> *The greatest value of a picture is when it forces us to notice what we never expected to see.*
> ~John W. Tukey, Exploratory Data Analysis

Data Visualization [DataViz] is the graphical presentation of multidimensional data with the goal of carrying out analyses and exploration of data to identify patterns, associations, trends, and so on. A graphical presentation of quantitative information always involves some aspect of relationships among the values. Good data visualization helps users explore and understand the patterns and trends in data and communicate that understanding to business users to help them make robust decisions by providing an effective representation of the underlying data being presented.

For now, DataViz practice is more in the realm of art than it is engineering, or even a scientific discipline. Most people who are responsible for analyzing data have never been trained to do this. Knowing how to use Excel or any other software that can be used to analyze data is not the same as knowing how to make sense of data. DataViz practice needs to define a body of knowledge and a set of rules and practices before questions such as the ones below can be answered with certainty:

- How to select the best graph to visually display a data set underneath.
- How to interpret a scatter plot or box-and-whiskers graph and when to use them.
- How to crate the charts and graphs not there to impress the users but to answer their questions.
- How to prioritize information with plots of trend patterns.

The basic difficulty that DataViz practitioners are facing is that there do not exit sufficient theories to describe and predict important phenomena of information visualization. Most fundamental questions are largely unanswered. Why does a certain visualization technique work better than another? How can we make a visualization tool better? How do viewers gain insights? The existing theoretical work is far from sufficient to help in answering these questions. The DataViz discipline is still in its inception and the biggest barrier to forming reasonable theories is a lack of empirical evidence.

Tables are used to present detailed [or summarized] levels of a data set, but just by looking at the data in the table, it's impossible to ascertain any trends or relationships. It is the graphs that enable visual communication of information about variables in the data sets, such as: trends, ranges, relationships, frequency distribution, comparisons, etc. The goal of a good graph is to display and communicate patterns and relationships of the underlying data with simple clarity. Additionally, looking at multiple graphs simultaneously [case for dashboard], can offer a new insight about data sets.

A DataViz practice was invented by Edward Tufte [Ref. 6] and he still rains in it. Stephen Few [Ref 7] followed the work of the great maser. Stephen Few [Ref. 8] is seen as an illuminator of Tufte's ideas as they pertain to DataVis applied to business. Edward Tufte and Stephen Few are often cited together.

The purpose of analytical displays of evidence is to assist in thinking. Consequently, when constructing displays of evidence, the question is: *"What are the thinking tasks that these displays are supposed to serve?"* The essential claim of Tufte's book *"Beautiful Evidences"* is that effective analytic design entails turning thinking principal into seeing principals. So if the thinking task is understanding causality, the task calls for design principal: *"Show causality."* If a thinking task is to answer a question and compare it with alternatives, the design principal is to *'Show comparisons.'"*

In the excellent book *"The Visual Display of Quantitative Information"*, Edward Tufte describes the value that good data graphics must have[1]. Good graphics must "*above all else, show the data*", and should:
- *Help the audience think about the important message(s) from the data, rather than about methodology (graphic design, the technology of graphic production etc), or something else*
- *Avoid distorting what the data have to say*
- *Present many numbers in a small space - but also emphasize the important numbers*
- *Make large data sets coherent, and encourage the audience to compare different pieces of data*

---

[1] Tufte (2001): *"The Visual Display of Quantitative Information"*

- *Reveal the data at several levels of detail, from a broad overview to the fine structure.*

Even before Tufte and Few, authors like William S. Cleveland [Ref. 9] had explored DataViz with paper and pencil. Messages arriving to our brains are predominantly [70%] delivered though our eyes. One of the leading experts in visual communication, Dr. Colin Ware, explains:

> *Why should we be interested in visualization? Because the human visual system is a pattern seeker of enormous power and subtlety. The eye and the visual cortex of the brain form a massively parallel processor that provides the highest-bandwidth channel into human cognitive centers. At higher levels of processing, perception and cognition are closely interrelated, which is the reason why the words 'understanding' and "seeing" are synonymous. However, the visual system has its own rules. We can easily see patterns presented in certain ways, but if they are presented in other ways, they become invisible...The more general point is that when data is presented in certain ways, the patterns can be readily perceived. If we can understand how perception works, our knowledge can be translated into rules for displaying information. Following perception based rules, we can present our data in such a way that the important and informative patterns stand out. If we disobey the rules, our data will be incomprehensible or misleading.* Information Visualization, Second Edition, Colin Ware, Morgan Kaufmann Publishers, 2004.

There are no, at this point in time, well-rounded data visualization theory. There are some general principles and from there a set of "best practices" can be derived. In the next section some basic data visualization principles using Excel charting techniques are demonstrated. The Excel charts can be effectively used to create powerful data visualization objects. But we emphasize, to becoming master of the trade, it's not the tool, it's the knowledge how to get the most out of the data, deeper understanding of what charts should be used and when is required and how to analyze and visually communicate quantitative information.

"Less is More"

> *Simplicity is the ultimate sophistication.*
> ~Leonardo da Vinci

This is the basic principle in DataViz and it can be traced back to Edward Tufte. In his book "*The Visual Display of Quantitative Information*" [Ref 6], Tufte refers to these concepts as "data-ink ratio", "data-density" or "chartjunk". "Data graphics should draw the viewer's attention to the scene and substance of the data" (Tufte); "The purpose of visualization is insight, not [pretty] pictures" (Shneiderman). Tufte brilliantly transposes this concept from Ludwig Mies van der Rohe's minimalism to the field of data visualization. The minimalistic approach is applicable to any field of human endeavour and, most certainly, to human communication. We are quoting a brilliant illumination of Tufte's "data-ink ratio", a basic principle of data visualization by Stephen Few. In his book "*Show Me the Numbers: Designing Tables and Graphs to Enlighten*", he writes:

> *The process of reducing the non-data ink involves two steps:*
> *1. Subtract unnecessary non-data ink*
> *2. De- emphasize and regularize the remaining non-data ink*

> *Subtract unnecessary non-data ink*
> *The process of subtracting unnecessary non-data ink involves asking the following question about each visual component: "Would the data suffer any loss of meaning impact if this were eliminated?" If the answer is "no", then get rid of it. Resist the temptation to keep things just because they're cute or because you worked so hard to create them. If they don't support the message, they don't serve so hard to create them. If they don't support the message, they don't serve the purpose of communication. As the author Antoine de Saint-Exupery suggests: "In anything at all, perfection is finally attained not when there is no longer anything to add, but when there is no longer anything to take away."* Show Me the Numbers: Designing Tables and Graphs to Enlighten", page 118 under heading: "Reduce the Non-Data ink"

The notion that "nothing" is an important "something" represents Tufte's concept of reducing "data-ink ratio". Increasing the amount of blank space is compensated by enhancing the importance of what is left. Indeed, when there is less, we appreciate more of what is left. In other words, the graphic should focus on the message(s) for the audience, and all visual clutter should be kept to a minimum.

What's Wrong With Pie Charts?

Pie charts are perhaps the most ubiquitous chart type; they can be found in newspapers, business reports and many other places. In spite of their widespread use, many data visualization writers like Edward Tufte, Stephen Few and Naomi Robins do not recommend the use of pie charts! Tufte, in *"The Visual Display of Quantitative Information"* [Ref. 6], wrote: "The only worse design than a pie chart is several of them". Stephen Few, in his highly regarded book *"Show me the Numbers"*, says: "I don't use pie charts, and I strongly recommend that you abandon them as well." Instead, "Dot Plots" or simple "Bar charts" are recommended as an excellent alternative to pie charts, as they show data positions along a common scale rather than relying on pie chart angles.

Here is the proof, by exhibiting an example, for the above statements:

Description	Surface area [km²]	Surface area in percent [%]
Saltwater	352,103,700	69.03
Freshwater	9,028,300	1.77
Farming land	44,682,307	8.76
Mountains	29,788,205	5.84
Snow-covered land	29,788,205	5.84
Dry land	29,788,205 km²	5.84
Land without topsoil	14,894,102	2.92

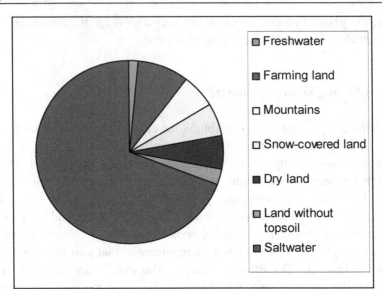

Fig 6.4 Pie Chart

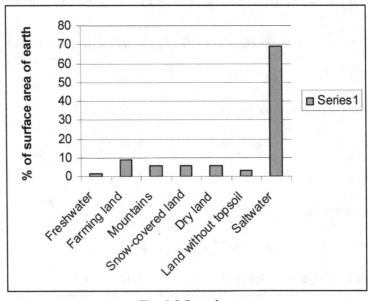

Fig 6.5 Bar chart

The simple rule for use of the Pie Charts is: if it has more that 3 slices, do not use Pie Charts, unless it's one of these:

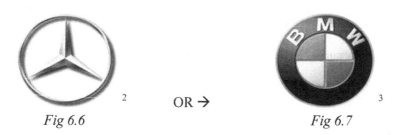

Fig 6.6 ²   OR →   Fig 6.7 ³

Fig 6.6   *Fig 6.7*

It is not with lightheartedness that we are displaying the above symbols of German "Gestalt" philosophy, but to present the functional beauty of the graph principle, another important attribute of data visualization.

The German word for design is "Gestaltung"; this term originated in the famous Bauhaus school, founded in 1919. The basic principle of Gestalt is to aspire to find the conceptual fit that resonates with the mind. The goal of Gestalt practitioners has been the relentless pursuit to find the most appropriate gestalt that benefits a need.

If the objects we are creating are going to inspire conceptual thinking and business insight [i.e. Business Intelligence], they must have aesthetic appeal. They must be in harmony with our user's minds!

Representing numbers

One of the oldest methods of representing numerical quantity is by tally marks. When counting a number of items, we add a vertical line "|" for each item. For example, the number three would be represented by three vertical lines "III", or the number fifteen by ["IIIII IIIII IIIII"]. Tally marks are still in use today, as this is an easy way to compare two or more numerical values. The size of a number is directly proportional to its length. This notation is supported by MS Excel. Using the Excel function: =REPT [text, number_to_tally], visually presents monthly sales as below:

---

[2] Trade Mark of Mercedes

Month	Sales	Tally Representation
January	33,000	|||||||||||||||||||||||||||
February	35,000	|||||||||||||||||||||||||||||
March	35,555	|||||||||||||||||||||||||||||
April	36,789	||||||||||||||||||||||||||||||
May	38,000	||||||||||||||||||||||||||||||
June	39,000	|||||||||||||||||||||||||||||||
July	40,657	|||||||||||||||||||||||||||||||
August	41,000	||||||||||||||||||||||||||||||||
September	42,000	||||||||||||||||||||||||||||||||
October	42,998	||||||||||||||||||||||||||||||||
November	43,000	|||||||||||||||||||||||||||||||||
December	50,000	|||||||||||||||||||||||||||||||||||||||

Improving on the above representation by changing the formula to display bar chart and the rank:

To show the bar chart: =REPT("█",D5/MAX($D$5:$D$12)*25)
And for rank: =RANK(D6,$D$5:$D$14,0)

Month	Sales	Sales Representation	Rank
January	33,000		7
February	32,000		8
March	35,555		5
April	36,789		4
May	38,000		3
June	39,000		2
July	35,000		6
August	41,000		1

Roman numerals are another way of representing numbers. The length of the roman number, as the tally counts, also represents its numerical value. The numerical value of the number 8 is represented by the tally count "IIIII III". The same number using Roman numerals is represented by "VIII". Modern Arabic numerals are irreplaceable for complex calculations, but may not be the best choice for visual representation. In fact, just by looking at the table above, it is obvious that a tally is superior to Arabic representation of total sales numbers. Where a simple comparison is required, graphic representation is a better choice, but where numerical values are used for numerical calculations, Arabic notation is superior.

But the story that we use to frame the numbers is also important, as studies have shown that the decisions are affected by how a story frames the number as oppose to numbers alone:

In a survey, when doctors ware told that the *mortality rate* for a certain operation is 6%, they hesitated to recommend it. On the other hand, when they ware told it had *survival rate* of 96% they ware more inclined to recommend it to their patients.

The facts presented in both cases ware correct. Framing the facts with *"survival rate"* triggered more optimistic story than the phrase *"mortality rate"*.

"War and Peace" by Tolstoy presented visually

The highlight point at Professor Edward Tufte's seminar *"Presenting Data and Information"* happens when he presents **Charles Joseph Minard**'s famous statistical graph of Napoleon's disastrous Russian campaign 1812 - 1813. [*"Carte figurative des pertes successives en hommes de l'Armée Française dans la campagne de Russie 1812-1813"*]. The map follows the French invasion and retreat from Russia in 1812. Tufte in his *"The Visual Display of Quantitative Information"* [p. 40] regarding, Charles Joseph Minard's 1861 thematic map of Napoleon's ill-fated march on Moscow refers to it as:

"It may well be the best statistical graphic ever drawn"

He uses it as a prime example in his seminars and also in his acclaimed 1983 book: *"The Visual Display of Quantitative Information"*. "This is", Tufte said:

"War and Peace as told by a visual Tolstoy."

Napoleon Bonaparte began his ill-fated 1812 invasion of the Russian Empire with 422,000 men. With each step further into Russian territory, more and more soldiers died or deserted. By the time it reached Moscow, Napoleon's army had dwindled to 100,000 men–already less than a quarter the size it had been at the start. During their disastrous retreat out of Russia, temperatures plunged to −37.5 °C. Nearly half the remaining survivors of the invasion were killed during the botched crossing of the Berezina River. Of the 422,000 men who set out on the invasion, barely 10,000 of them returned alive.

All this information is readily visible in the above chart, created by Minard combined both a map of the campaign and a visual representation of the number of men remaining in Napoleon's doomed army. The thickness of the line is proportional to the number of men in the army (one millimeter equaling 10,000 men), with the gray [original color map is in beige] section representing the offensive toward Moscow, and the black line the retreat. Below, Minard also included a second chart showing the temperature on various days during the retreat.

Minard includes a description above his chart, but it is almost completely unnecessary; all the pertinent information is readily apparent from a close examination of the chart itself. Minard was a master at the production of maps such as these that combined tremendous amounts of data with geographic representations. The most striking feature of the chart is the thinning line of soldiers; the background of the map showing the cities and rivers the army traversed on its way into and out of Russia.

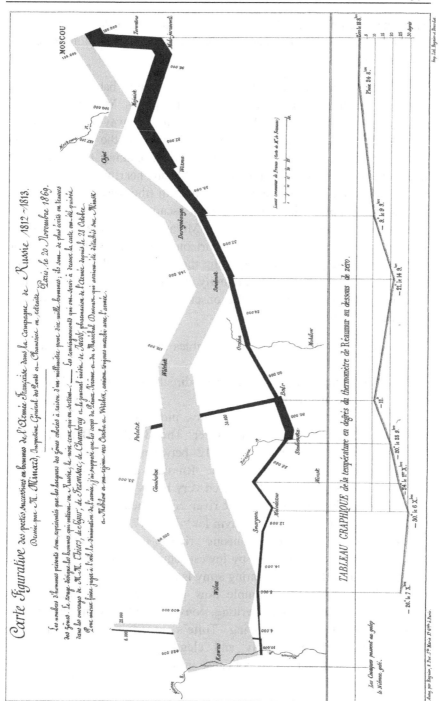

Fig 6.8

This chart demonstrates how, with good planning a design, maps can operate in concert with many other types of information to create stunning displays of information.

A frustrated Napoleon had little choice but to return back to the part of Europe he controlled for food, shelter, and supplies. Minard now traces the remnants of the "Grande Armee" as it makes its way back toward the Neiman River. In doing so, the parallel tracks of the advancing and retreating army are set next to one another, making the continuing deterioration of the army all the more visible and heart wrenching. As the army slowly made its way across barren earth (the Russians had burned food along this path while blocking other escape paths), one of the worst winters in recent memory set in. Minard tracks the plummeting temperature against this trek on a horizontal axis at the bottom of the page, even more profoundly capturing the dire straits that the retreating army found itself in. Not surprisingly, the pitiful band of troops that returned from Russia marked the onset of the collapse of Napoleon's Continental Empire.

The graph [map] displays several variables in a single two-dimensional space like:

- The army's size, location and direction, showing where units split off and rejoined
- The declining size of the army, most striking example is the crossing of the Berezina River. The **Battle of Berezina** took place November 26–29, 1812 between the retreating French army of Napoleon and the Russian armies under Mikhail Kutuzov. The French suffered very heavy losses but managed to avoid the trap and cross the river. Since then "*Bérézina*" has been used in French as a synonym for "disaster"
- The mild weather causing rivers to overflow its banks immediately followed by lowest temperatures during the retreat and its effect on retreating army is obvious from the map.
- The Map is an example how good data, when treated with respect, like a good writing, convey beautifully and clearly the meaning. Thus proving Tufte's assertion that data when presented respectfully and elegantly is not confusing, but clarifying and it does not cause information overload,

Seeking Relationships

Relating data or finding accurate, convenient and useful representations of data, involves, first determining   the nature and structure of representation [E.g., linear regression] and then deciding how   to quantify and compare the two [actual data with regression model] different representations [E.g., sum of squared errors]. Once the model has been built and validated, it can be used to make predictions. Along with prediction, there should be same indication of the confidence of the prediction.

The most popular model for making prediction is multiple linear regression model. This model is used to fit a linear relationship between a quantitative dependant variable Y [also known as outcome] and a set of predictors X_1, X_2, ...X_p [referred as independent input variables or regressors]. The basic assumption is that the following relation holds:
$$Y = \alpha + aX_1 + bX_2 + \ldots + pX_n + \beta$$

EXAMPLE of Linear Regression with one variable: $y = a + bx$

Predictor variable = x (i.e. income), Response variable = y (spending)

Score is defined as the sum of squared errors

Need to find a and b such that $y = a + bx$

Assuming a set of numbers

Y	X
1	3
8	9
11	11
4	5
3	2
12	
15	

Scatterplot is created first to identify whether any relationship exists between the two variables. The two variables are plotted on the x- and y-axes. Each point is displayed on the scatterplot is a single observation. The scatterplot allows seeing the type of relationship that may exist between two variables.

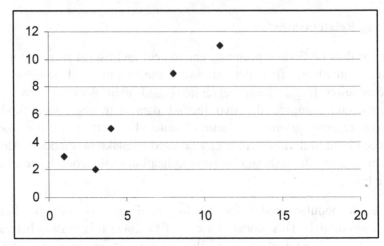

Fig. 6.9 Scatterplot

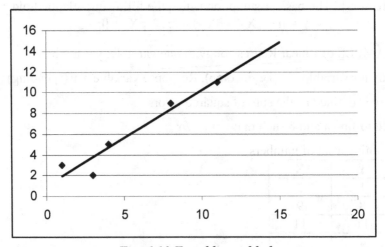

Fig. 6.10 Trend line added

Values of "a" and "b" [regression coefficients] can be calculated or derived from the trend line. The second method – deriving values for "a" and "b" from the trend line is used here. The point at which the trend line intercepts with the y-axis is a value of "a". The slope of the line is the value for "b" [b = y/x]. The derived formula for the optimal regression line [relationship between the income and spending], for the set on numbers in the table above is:

$y = 0.8 + 1.4x$

Comparing response values with the predicted value it is obvious that there are differences. In the case of perfect prediction model, all the actual points would lie on the prediction line [trend line]. More practical assumption is a good model, which have the actual points close to the prediction line. Confidence of the predictive accuracy is of a good model is described below:

The error or residual is defined as the difference between the actual response values and predicted values. "Correlation coefficient" is calculated based on the actual and predicted value. The resulting value of the correlation coefficient is between -1 and +1, where strong positive correlations are signified with values closer to +1, strong negative relationship are closer to -1, value of close to zero indicates absence of any relationship. When the values are squared the resulting range will be between 0 and 1. Model accuracy improves as correlation coefficient or squared correlation coefficient values get closer to 1 [1 corresponds to 100% accuracy].

Indeed, the notion of prototyping is closely related to validating user requirements! In software, nothing is visible until it is done, meaning that a change of mind by users [typically non-programmers] may be too late. Software products are typically easy to use only if they are designed by programmers for other programmers and not for normal human beings. Or conversely and in accordance with Grudin's Law:

> *"If the person responsible for using the system does not benefit, the system is doomed to failure."*
> ~ Jonathan Grudin, professor, Information & Computer Science Department, University of California, Irvine

Typically, users cannot design and program their own BI solutions and are dependant on their IT colleagues. This may change in the future as major MBA program around the world are introducing art and design in their curriculum. Bridging the Grudin's Law necessitates, for now, strong collaboration between the two groups [designers and users], from the start of the program the final solution. Designers must continuously validate that they are building the right product, as it is always cheaper to build it right the first time than to build it over and over again, especially in this case, when providing users with cognitive objects! This necessitates a human-centered design process implemented with rapid prototyping and an iterative process to solve such a complex problem.

The concept behind a prototype is that users cannot tell what it is they want, but they can tell quickly what they don't want when they see it. A user-centered design process requires going to the places where the domain experts [users] work with the purpose of understanding people and being able to extract innovative ideas. It involves looking for the latent need, a need that has not been expressed in any way. On the pragmatic level, it means sitting down with the users, watching them use the reports, looking out for when they are in trouble or even frustrated, looking for the moments when they are having a good time, and amplifying the good experience in the cognitive objects built for the users.

Excel is an amazing tool that could be used to build the rapid prototype. It is the best tool for executive dashboard prototyping because of its flexibility and low development costs. Creating a fully functional prototype dashboard does not require a development programmer to do and it should not be done by the development team, as this is a QA tool used for continues requirement validation and prone to frequent changes. It should be kept out of the rigorous development version control process. The tool is used for quick modeling and user feedback should be available in a matter of days. Every time you need a dashboard project, think of a prototype in Excel.

Excel is a very powerful visualization tool, tool that can be effectively used with no, or very little VBA programming. Below are simple visualization graphs:

	Jan	Feb	Mar	Apr	May	Jun	Jul	Aug	Sep	Oct	Nov	Dec
Target	30,029	31,187	28,542	34,006	36,465	35,001	39,270	39,543	33,810	39,005	39,000	39,000
Revenue	30,666	31,685	29,342	33,773	37,376	37,143	40,184	40,381	35,600	38,170	37,880	41,770
Variance to Target	$637	$498	$800	($233)	$911	$2,142	$914	$838	$1,790	($834)	($1,120)	$2,770

	2009	2010	2011
J	88	135	199
F	62	110	190
M	65	109	186
A	59	99	171
M	99	140	198
J	80	115	190
J	88	130	170
A	90	149	169
S	115	150	159
O	119	140	166
N	109	135	162
D	101	119	171

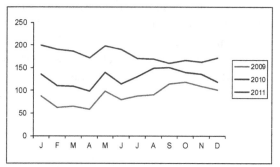

Trending graph

	Jan	Feb	Mar	Apr	May	Jun	Jul	Aug	Sep	Oct
Target	31,021	32,187	29,542	33,006	37,465	35,089	38,270	37,543	35,810	37,465
Revenue	30,685	31,685	30,342	32,773	35,376	36,143	39,184	37,381	36,600	37,170

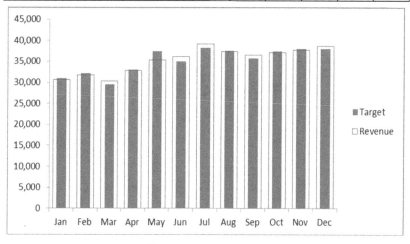

Revenue vs. Target

Many Excel Users rely on Excel's default charts types and settings to produce their charts and graphs without regard for data visualization principles. Default Excel charts are not effective data visualization tools. By combining data visualization principles and advanced Excel charting techniques, Excel charts can be made into powerful data visualization tools. An Excel prototype can be used for rapid and iterative development. It is easy to create multiple alternative user interfaces, get feedback from users or find design flaws. The prototype is used to continuously validate user requirements. The final version of the prototype is used for creating specifications for the IT department to construct the WD application.

The goal when designing the dashboard is to craft a visual story that is in harmony with all charts and graphs and resonate seamlessly with the user's thinking, Dashboards can tell a much richer story than a simple chart. If we think of one chart as a short story, then a dashboard is a novel where characters are fully developed.

Testing – Mostly about Verification

> *"Program testing can be used to show the presence of bugs, but never to show their absence"*
> ~ Dr. Edsger W. Dijkstra

Most complex software quality assurance activities are centered around testing. The general aim of testing is to affirm the quality of software systems by systematically exercising the software in carefully controlled circumstances. A good test is one that has a high probability of finding an as yet undiscovered defect. A successful test is one that uncovers an undiscovered defect.

Software testing maturity model is based on systematic CMM [Capability Maturity Model] methodology for continuous process improvement. Each testing maturity level has its set of processes and practices and are directly related to CMM laves. The purpose is to provide testing framework for assessing the maturity of the test processes in an organization, providing targets for improving testing maturity.

This testing strategy defines five levels of software testing maturity, as follows:

Level 1 – Initial: at this level, an organization is using ad-hoc methods for testing, so results are not repeatable and there is no quality standard. This looks more like debugging than testing. The objective of this kind of testing is to prove that the application has no defects.

Level 2 – Definition: at this level, testing is defined as a process, so there might be test strategies, test plans and test cases, based on requirements. Testing does not start until products are completed, so the aim of testing is to compare products against requirements.

Level 3 – Integration: at this level, testing is integrated into a software life cycle, e.g. the V-model. The need for testing is based on risk management and the testing is carried out with some independence from the development area.

Level 4 - Management and Measurement: at this level, testing activities take place at all stages of the life cycle, including reviews of requirements and designs. Quality criteria are agreed upon for all products of an organization [internal and external].

Level 5 – Optimization: at this level the testing process itself is tested and improved with each iteration. This is typically achieved with tool support, and also introduces aims such as defect prevention throughout the life cycle, rather than defect detection [zero defects][1].

Testing has been widely used as a way to help software engineers develop high-quality systems. The maturation process through which the testing techniques for software have evolved, from an ad hoc, intuitive process to an organized systematic software engineering discipline. Part of a strategy document is to identify the type of testing required. The primary focus in this section is to describe each test identified in the testing strategy with the focus on the verification that correct data is transferred into the Data Warehouse.

Limitation of testing: Any non-trivial system will have defects. We cannot know in advance which inputs and in which sequence of execution will cause system to fail. Outcome may depend on internal state of the system when a particular code is executed, If a particular function returns a correct result, we cannot guaranty that function will always return correct results. And it is impossible to test a function for all possible vales. Testing is not a substitute for defect prevention software engineering practises. Testing should be complemented by other [inspection for example] software engineering practises.

Data Quality
The most important assumption to make when discussing data quality is that all conclusions of data analysis are subject to qualifications about the quality of the data. The computer acronym GIGO, standing for Garbage In, Garbage Out, applies just as much here as elsewhere, and data analysts of whatever flavor, cannot perform miracles and extract gems from rubbish.

Here is a question: "Why is dirty data the norm rather than the exception?" The answer is that dirty data is the norm simply because we make the wrong assumption: let's assume the data is always correct, shall we? No, we shall not, as the norm is just the opposite unless something is done about it. This section deals with causes and offers some solutions to the "dirty data" phenomenon.

---

[1] CrossTalk magazine Aug. 1996: Ilene Burnstein, Taratip Suwannasart and C.R. Carlson: Developing a Testing Maturity Model

Data quality is one of the most important and most difficult aspects of an enterprise's data management efforts. The fundamental principle that any data analyst needs to know about data quality is that data has to be suitable for its intended use. It is important to understand the context in which the data is being used and that user satisfaction is tied directly to data quality. The foundation of data quality starts with the conformance of data, data types or data attribute value for that domains.

Before we start verifying data quality, we have to know how the ETL system handles data rejection, substitution, correction and notification without modifying data. To ensure success in testing data quality, include as many data scenarios as possible. Typically, data quality rules are defined during design, for example:

- Reject the record non-numeric data if it is a numeric field
- All source data that is expected to get loaded into target actually gets loaded [compare counts between source and target and use data profiling tools]
- All fields are loaded with full contents − i.e. no data field is truncated while transforming
- No duplicates are loaded
- Substitute null if a decimal field has non-numeric data
- Aggregations take place in the target properly
- Verify and correct the state field if necessary based on the ZIP code
- Data integrity constraints are properly taken care of
- Verify results based on the business rules [if, for example the business rule is:
  "If there is no match found in the look-up table, then load it, but report the error to a user in the error log/report"]

Depending on the data quality rules of the application being tested, scenarios to test might include null key values, duplicate records in source data and invalid data types in fields [e.g., alphabetic characters in a decimal field]. Review the detailed test scenarios with business users and technical designers to ensure that everyone is in agreement. Data quality rules applied to the data will usually be invisible to the users once the application is in production; users will only see what is loaded to the database. For this reason, it is important to ensure that what is done with invalid data is reported to users. These data quality reports present valuable data that sometimes reveals systematic issues with source data. For the purpose of continuous quality improvement efforts, it may be beneficial to create a quality database and populate the "before" data in this database for users and the QA team to view, in order to be able to take appropriate actions.

In order to prevent the propagation of errors, data quality must be fixed at the source.   Early in the testing phase, data quality should be instrumented with a "quality filter" to detect and report quality.   New quality filters should be added incrementally.   Data quality should be monitored and continuously improved by employing six sigma methodologies.

Metadata
A neglected but very important component of any database, but even more so of the DW, is metadata [data about data].   Metadata and data heredity provide the business definitions of the data, the technical specifications of the data, and a process that shows how the data has been processed, what business rules are applied to the data, and what validation criteria were applied to the data.   In the new world of DW, metadata has taken a higher level of importance and it has become imperative for it to be part of any DW.   Without it, data analysts, programmers and others would be poking into DW forever without any guarantee of success.   Metadata is like an index into DW as it stores the following information:

- Structure of the data from the programmer's point of view
- Structure of the data from the data analyst's perspective
- Sources of data feeding DW
- Business rules for data transformation
- Data model
- Relationships of data
- History of extractions

Unit testing:

"A test that reveals a bug has succeeded, not failed."
~ Dr. *Boris Beizer* - Software testing techniques

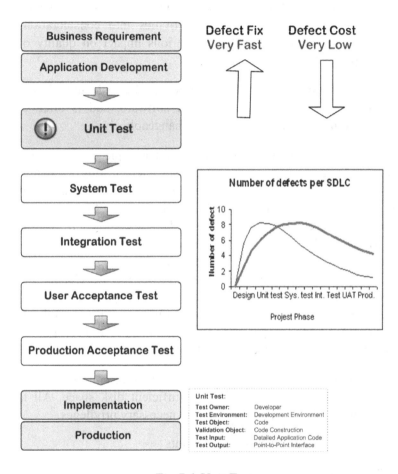

Fig 7.1 Unit Test

Unit test starts after the design reviews and code inspections are completed. Depending on the outcome of these 'pre-unit test' activities, project defect trend can take one of the two lines on the 'Number of Defects per SDLC' graph on the previous page. If the design inspections and code reviews are done correctly [as prescribed by CMMI standards] number of problem potential defects will be uncovered early. In fact, some of the most difficult, i.e. structural problem with the code can only be uncovered during the design review or code inspection stage. It is sometimes difficult or even impossible to create conditions in test environment to uncover these kinds of problems. If the defects are not found out during unit testing we will be finding and fixing them in system testing, or worse, in the field, where the cost will be orders of magnitude higher But, rest assured that in the dynamic of the production environment they will show up sooner rather than later. The testing strategy must take these important activities into consideration and strictly enforce them as they have a serious impact on quality of the delivered product.

Unit Testing is done at the lowest level. It tests the basic unit of software, which is the smallest testable piece of software, and is often called "unit", "module", or "component" interchangeably.

Unit Testing is the lowest level [i.e., a unit is a piece of source code] using a coverage tool; as a minimum, sufficient testing to assure that every source statement has been executed at least once. It may be necessary to test each branch [both TRUE and FALSE] to assure the 100% branch coverage of a unit of code. Unit testing is done to find unit bugs, which is the second highest frequency kind of bug. Unit testing refers to the testing of discrete program modules and scripts. This has traditionally and logically been the task of the developer. Every developer must test her program modules and scripts individually. This type of testing is known as a white-box testing. The purpose of this test is to ensure the module or unit of code is coded as per agreed upon design specifications. The developer should focus on the following:

1. That inbound and outbound directory structures are created properly with appropriate permissions and sufficient disk space. All tables used during the ETL are present with necessary privileges.

2. The ETL routines give expected results

3. Data Cleansing
   - ETL transformation logics from source to target work as specified in designed documentation
   - Boundary conditions are satisfied [e.g. check for date fields with month of February, leap year dates, zero or less years of age, etc.]
   - Surrogate keys have been created as Primary key as applicable [Conversely neither Natural nor Compost keys should not be used as Primary key]. Database indexes and referential integrity

constraints can help avoid some these problems, but not all data is within a relational database. Some of the relations cross databases and technologies, but even within databases not all dat can afford to implement referential integrity constraints.

- Appropriate fields are initiated with NULL values where expected
- Log of rejects is created where applicable, with sufficient details
- All error recovery methods and/or routines are verified. Sometimes, during testing, it is not possible to create condition, which would invoke an error recovery method. The code involved in this particular situation must be reviewed with peers or appropriate knowledgeable team members.
- Auditing for any update, delete or insert into DW is done properly

4. That the data loaded into the target is complete:
  - All source data that is expected to get loaded into target actually got loaded.
    Pre and post row counts and checksums work wonders here.
  - All fields are loaded with full content– i.e. no data field is truncated while transforming
  - No duplicates are loaded
  - Aggregations take place in the target properly
  - Data integrity constraints are properly taken care of

Unit Test Automation

Automated unit test is a piece of code, written by a programmer that exercises another piece of functional code. Unit testing is performed to verify that a piece of code does what the programmer thinks it should do. Unit testing is relatively easy practice to adapt, but there is a set of guidelines that has to be followed to make it affective. This is demonstrated here:

There exists an open source unit testing framework, initially develop by Steven Feuerstein, modeled after Junit and xUnit frameworks. Commercial version is available from Quest and Steven Feuerstein.
Here the demonstration of the steps involved in unit testing Pl/SQL procedure:
1. Select the table and PL/SQL procedure to test.
2. Create the Unit Testing Repository.
3. Create the Unit Test.
4. Run Unit Test
5. Create and run and run an Exception Unit Test
6. Create a Unit Test Suite
7. Run Unit Test Suite and produce the report.

Unit Test Automation Example

Quest tool Code Tester for Oracle is used to demonstrate unit test process automation. In this example VENDOR_RATING field has become mandatory for all active accounts. If the VENDOR_RATING value for an active member is missing default value of 100 is inserted. For the vendors where there is a value in the field VENDOR_RATING, the field is not updated.

Create the VENDOR_INFO Table with sample data:

```
CREATE TABLE vendor_info (vendor_id NUMBER PRIMARY KEY,
vendor_name VARCHAR2(30),  active_flag CHAR(1), sale_amt
NUMBER, vendor_rating NUMBER, modified_date DATE);

INSERT INTO vendor_info VALUES (10002001, 'ABC Company',
'Y', 98000, 101, '01-jan-10');

INSERT INTO vendor_info VALUES (10002003, 'TTT INC', 'Y',
200000, null, '01-jan-10');

INSERT INTO vendor_info VALUES (10002021, 'DDD TECH', 'Y',
1000, 91, '01-jan-10');

INSERT INTO vendor_info VALUES (10002035, 'CC INC', 'Y', 0,
null, '01-jan-10');
```

```
INSERT INTO vendor_info VALUES (10002061, 'R BANK', 'N',
5000, null, '01-jan-10');

INSERT INTO vendor_info VALUES (10002062, 'B INC', 'Y',
39000, 80, '01-jan-10');

INSERT INTO vendor_info VALUES (10002094, 'D Computer INC',
'Y', 2000, null, '01-jan-10');

INSERT INTO vendor_info VALUES (10002097, 'Tom INC', 'N',
2000, 70, '01-jan-10');
```

VENDOR_ID	VENDOR_NAME	ACTIVE_FLAG	SALE_AMT	VENDOR_RATING	MODIFIED_DATE
10002001	ABC Company	Y	98000	101	1/1/2010
10002003	TTT INC	Y	200000		1/1/2010
10002021	DDD TECH	Y	1000	91	1/1/2010
10002035	CC INC	Y	0		1/1/2010
10002061	R BANK	N	5000		1/1/2010
10002062	B INC	Y	39000	80	1/1/2010
10002094	D Computer INC	Y	2000		1/1/2010
10002097	Tom INC	N	2000	70	1/1/2010

Fig. 7.2

Create the RATE_UPDATE Procedure:

```
CREATE OR REPLACE PROCEDURE rate_update (ven_id NUMBER)
IS
    v_rate    REAL;
    v_active CHAR (1);
BEGIN
    SELECT vendor_rating
       INTO v_rate
       FROM vendor_info
     WHERE vendor_id = ven_id;
    SELECT UPPER (active_flag)
       INTO v_active
       FROM vendor_info
     WHERE vendor_id = ven_id;
    IF v_rate IS NULL AND v_active = 'Y'
    THEN
        UPDATE vendor_info
          SET vendor_rating = 100
        WHERE vendor_id = ven_id;
    END IF;
END;
```

Screen-print of the stored procedure in the TOAD® [Quest Software]:

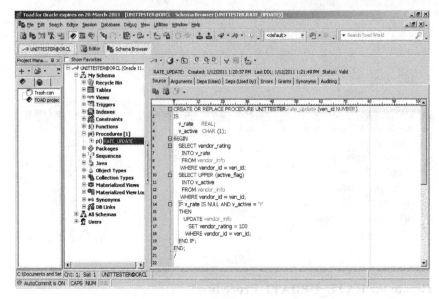

Fig. 7.3

```
CREATE OR REPLACE PROCEDURE UNITTESTER.rate_update (ven_id NUMBER)
IS
    v_rate    REAL;
    v_active  CHAR (1);
BEGIN
    SELECT vendor_rating
     INTO v_rate
     FROM vendor_info
    WHERE vendor_id = ven_id;
    SELECT UPPER (active_flag)
     INTO v_active
     FROM vendor_info
    WHERE vendor_id = ven_id;
    IF v_rate IS NULL AND v_active = 'Y'
    THEN
       UPDATE vendor_info
         SET vendor_rating = 100
        WHERE vendor_id = ven_id;
    END IF;
END;
/
```

Fig 7.4

Unit Testing Phase:

Create Test Matrix:

Test Case #	Expected Results	Test Data	Vendor ID
1	Insert the default vender rate	Active account VENDOR_RATING Null	'10002003'
2	The record is not be updated	Active account VENDOR_RATING not Null	'10002001'
3	The record is not be updated	Non-active account VENDOR_RATING Null	'10002061'
4	The record is not be updated	Non-active account VENDOR_RATING Not Null	'10002097'

Fig 7.5

Test Sample Tool: Quest Code Tester for Oracle – Beta Version 2.0:

Fig. 7.6

Step 1: Login to the Quest Code Tester for Oracle:

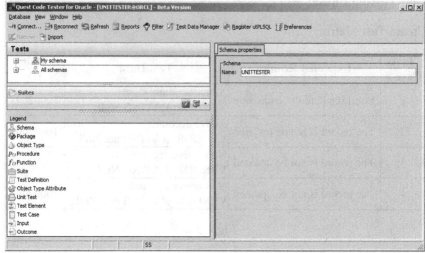

Fig. 7.7

Step 2: Choose the test procedure.

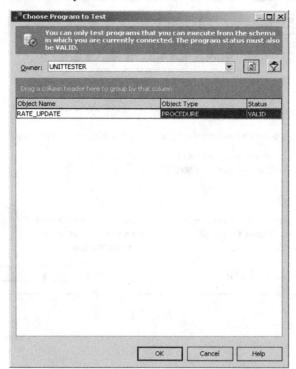

Fig. 7.8

Step 3: Insert the test cases.

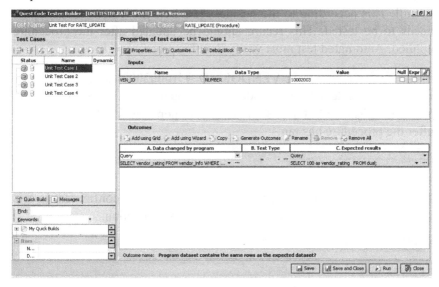

Fig 7.9

For example, Test Case 1- Vendor ID = '10002003':

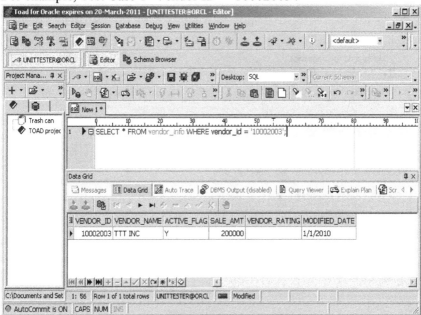

Fig 7.10

The ACTIV_FLAG value is set to 'Y' and the VENDOR_RATING is set to null. This according to the business requirement is the condition that needs to be tested. This record should be updated after the procedure applied. Automated test must verify that this process works.

In the example above for VENDOR_ID = '10002003' is the input value in the Test Builder of Quest Code Tester. The outcome value for the field VENDOR_RATING for this record should populate a default value 100 after the procedure has ran.

```
SELECT vendor_rating FROM vendor_info WHERE vendor_id =
'10002003';
```

Also make sure the related properties have been fully configured. In this case, for example, rollback is required after the execution.

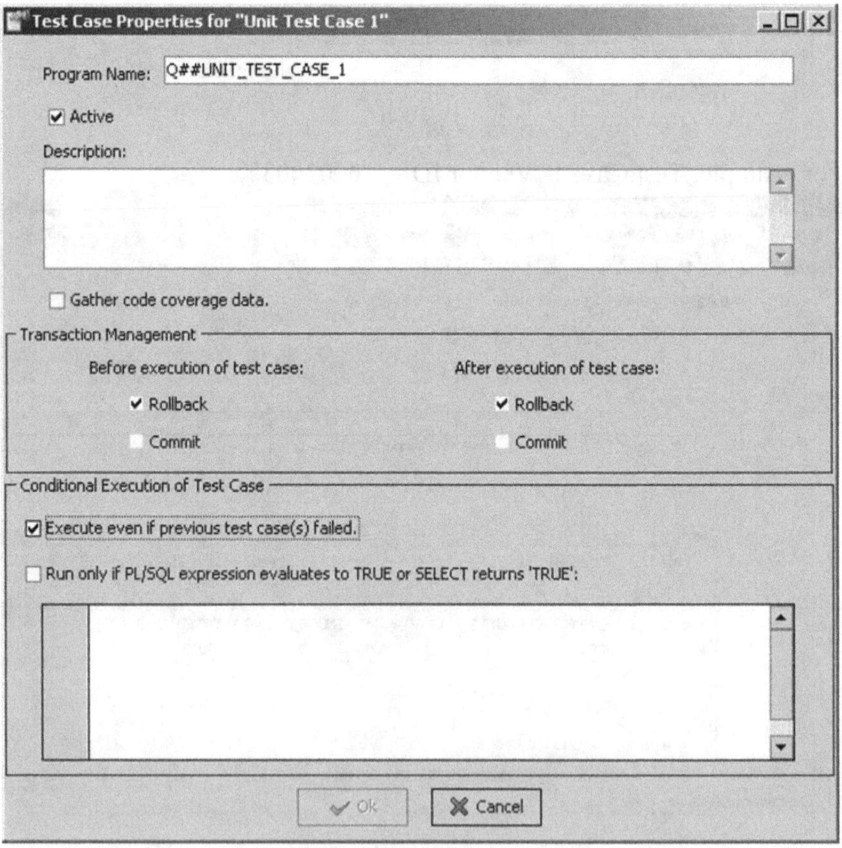

Fig 7.11

Step 4: Run all the test cases in Quest Code Tester.

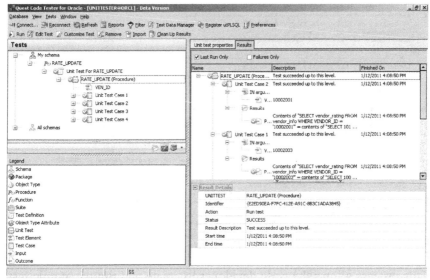

Fig 7.12

Step 5: Create Test Report.

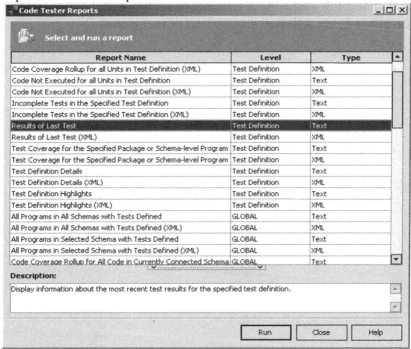

Fig. 7.13

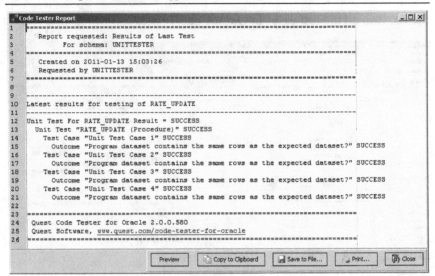

Fig 7.14

This process allows the code developer to create and run the atomic test case during code development and at the same time accumulating full Test Suite the in the Code Tester for Oracle tool. At the end of development full Test Suite is executed to confirm that the code is defect free, before being delivered it to the next project phase.

System Testing:

> *"Testers are people who realize that things can be different!"*
> ~ Jerry Weinberg

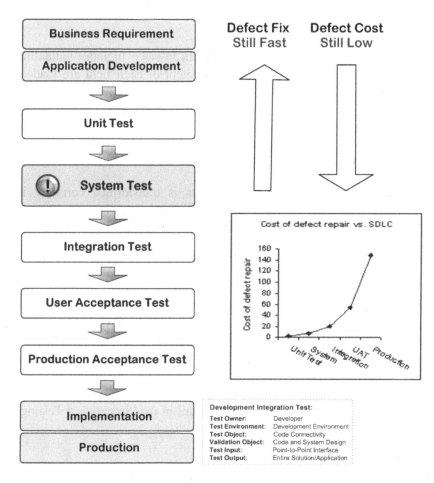

Fig 7.15 System Testing

Typically, the scope of system testing only includes testing within the ETL application. The endpoints for system testing are the input and output of the ETL code being tested. The validated design document is the Bible and the entire set of test cases is directly based upon it. The functionality of the application is tested; usually it is done using the black-box test methodology. The major challenge here is the preparation of intelligently designed input test dataset. This can uncover most of the problems in the ETL application quickly. Production-like data should be used wherever possible. We must select a test data to verify for all possible combinations of input and specifically check for errors and exceptions. Knowledge of the business process and business rule for data transformations are essential here, as we must be able to relate the results functionally and not just verify the code.

The testing team must verify for:
Data completeness or conversely No **Data losses** – Includes validating that all records, all fields and the full content of each field are loaded. Strategies to consider include:
- Record counts between source data, data loaded to the warehouse and rejected records should be reconciled
- Negative scenarios are also validated. [Verifying for special characters, etc.]

Correct data transformation - all data is transformed correctly in accordance with business rules as per design specifications. This could be simple 'one to one' or complex data transformation.

Data aggregations– match aggregated data against staging tables and/or ODS

Data quality - the ETL application correctly rejects and substitute default values, corrects or ignores and reports on invalid data. ETL testing revolves around the data, that is why is important to achieve a degree of excellence in data accuracy. We know we have achieved quality when we successfully fulfill customers' requirements. Data accuracy will deliver a value for our customer.
We will discuss this subject in further detail in the "Integration Testing" section.

Data Completeness
The most basic test of data completeness must be executed first. Objective of this test is to verify that all the expected data has been loaded into the data warehouse. This includes validating that all records, all fields and the full contents of each field are loaded. Testing scenario to consider in this verification includes:

- Verifying the count of record in the source data and data loaded to the warehouse plus the rejected records are equal.

- Comparing the unique values of key fields between source data and destination data [data loaded in the warehouse]. This technique may points out a variety of possible data errors without having to do full validation on all fields.
- Utilizing a data profiling tool or SQL filters that show the range and value distributions of fields in a data set. This can be used during testing and in production to compare source and target data sets and point out any data anomalies from source systems that may be overlooked if the data movement is correct
- Populating the full content of each field to validate the contents of each field ensuring that no truncation occurs at any step in the process. For example, if the source data field is a string 30, then make sure to test it with 30 characters.
- Testing the boundaries of each field to find any database limitations. For example, if a numeric field is defined as an integer whose length three [2] decimal number it must include all values between 9 and 99. If the field is defined date fields it must include the entire range of valid dates.

Log of bad records should be kept in an log table for audit and continues process improvement purposes. This concept should be carried over into production environment.

Data Transformation
Verifying that data is transformed correctly based on business rules can be the most complex part of testing an ETL application with significant transformation logic. Data transformation is mostly the application of business rules that need to be applied to the data. In data warehousing, this may include the computation or derivation of new values from existing data (such as profitability from cost and revenue), normalization or denormalization, filtering, classification, aggregation, or summarization of the data, etc.

One typical method of verification is to pick some sample records and compare by eyeballing to first validate data transformations manually. This is used as initial test step to do rudimentary verification, but it is a manual testing and requires full understanding the ETL logic. A combination of automated data profiling and automated data movement verification is a better long-term strategy.

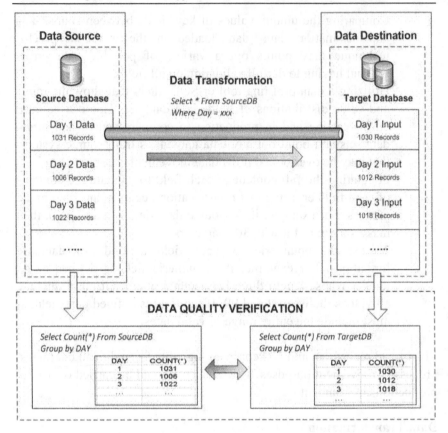

Fig 7.16

Here are some simple test scenarios that can be easily employed to validate and verify data movement within the DW:

- Create sets of scenarios of input data and expected results and validate these with the business customer. This is a good requirements elicitation exercise during design and can also be used during testing.
- Create test data for all the above scenarios. Elicit the help of an ETL developer to automate the process of populating data sets with the scenarios all possible data, making them test driven scenarios.
- Utilize data profiling results to compare the range and distribution of values in each field between source and target data.
- Verify correct processing of ETL-generated fields such as default values or surrogate keys.

- Verify that data types in the warehouse are the same as specified in the design and the data model.
- Scenarios that test referential integrity between tables. For example, what happens when the data contains foreign key values that are not in the parent table? Test how orphaned child records are handled.

Enterprise Integration Testing of DW

> *"To find the bugs that customers see - that are important to customers - you need to write tests that cross functional areas by mimicking typical user tasks. This type of testing is called scenario testing, task-based testing, or use-case testing."*
>
> ~ Brian Marick

The purpose of integration testing is to show how the DW application works from an end-to-end perspective. We must consider the compatibility of the DW application with all upstream and downstream flows, assuring data integrity across the flow. Following our notion of Validation & Verification, the first level of integration testing begins with validation of the warehouse data model. All subsequent testing is based on understanding this model and indeed, each element in the model and how it contributes to achieving specific business objectives in the BI area.

Validation
Granularity refers to the level of detail of the data stored in fact tables in a data warehouse. High granularity refers to data that is at or near the transaction level. Data that is at the transaction level is usually referred to as atomic level data. Low granularity refers to data that is summarized or aggregated, usually from the atomic level data. Summarized data can be lightly summarized data, as in daily or weekly summaries, or highly summarized data, such as yearly averages and totals.

Granularity is the single most important aspect to consider when designing a data warehouse. Data should be captured at its lowest, most atomic grain. Atomic data is highly dimensional. It is always possible to create higher level grain by aggregating atomic level data for a business process that may require it. The more detail there is in the fact table, the higher its granularity and vice versa. The consequence of the higher granularity of a fact table is that more rows will be required to store it. Let us look at an example to illustrate this important point:

We are going to take a look at a small data warehouse with a single fact [Sales] and three dimensions [Time, Organization and Product]. The fact table contains three metrics [Unit Price, Units Sold and Total Sale Amount]. The Time dimension consists of four hierarchical elements [Year, Quarter, Month and Day]. The Organization dimension consists of three hierarchical elements [Region, District and Store]. The Product dimension consists of two hierarchical elements [Product Family and SKU].

The metrics in the Sales fact table must be stored at some intersection of the dimensions of Time, Organization and Product. Therefore, in this data warehouse, the highest granularity that we can store Sales metrics in is in Day/Store/SKU [as these are the lowest level in each dimensional hierarchy]. Conversely, the lowest granularity that we can aggregate Sales metrics to in this data warehouse is by Year/Region/Product Family [as those are the highest levels in each dimensional hierarchy]. We may also [for a variety of performance reasons] choose to store Sales metrics at some intermediate level of granularity, such as by Month/District/SKU. In other words, summarized data can be lightly summarized, as in daily or weekly summaries, or highly summarized, as in yearly averages and totals.

The question we need to answer is: which level of granularity is right? The answer to this question depends on what business question this data warehouse needs to answer. Quite often, we choose to have fact tables with a high level and low [aggregated] level of granularity. That is why the question of right granularity for a data warehouse is an important issue to ponder upon when designing DW.

Storing data at the highest level of granularity appears not to be a bad idea. The data can be reshaped to meet different needs – of the finance department, of the marketing department, of the sales department, and so forth. Granular data can be summarized, aggregated, broken down into many different subsets and so forth. There are indeed many good reasons for storing data in the data warehouse at the lowest level of granularity. Many companies start out by doing just that, building a data warehouse at the highest level of granularity, and find that their true requirements and budget make building this Enterprise DW model impractical. Only at that time do they start evaluating how the system is going to be used. What kind of reporting will be generated off it? It may also not be a bad idea to evaluate a business vision and the users' expectations. Take all of this into account and build the database for efficiency too.

The impact of the granularity of a data warehouse is profound and it has to be carefully evaluated as it affects the size and the performance of the DW. At the same time, use of DW in the future has to be taken into account, as the implemented granularity is hard to change but it may impede the needs of business in the future.

Potential Problems with Fact Tables

The purpose of the fact table is to be able to easily get useful information out of the data. In order to do this, any complex math or unusual query requirements should be avoided. The easiest way to do this is to make sure the actual measures, the numbers, are additive across all rows.

The additive property of the numbers across the rows is a most precious gift that must always be preserved in EDW. An accurate summary of data is of paramount importance in OLAP. DW is conceptualized around facts and dimensions tables. Facts contain a measure of interest and dimensions have attributes used to select and aggregate measure of interest. Classification attributes are modeled in forms of hierarchies. For example, a classification of hierarchy of time detention shows that measure in the fact table can be aggregated from the lowest level of granularity - seconds - to progressively higher levels - hours and days. As an example, the density of telephone traffic may be aggregated from density of traffic per second to daily density of telephone traffic. Aggregation is usually done along multiple dimensions. We may want to know the daily density of traffic in Toronto for the year 2008. Note that the Date dimension may be a separate dimension. It is important to understand that measure can be additive across all dimensions, one dimension, or not additive across any dimension. The first step in ensuring accuracy of aggregation is to recognize when and how aggregation operation can be used.

Some examples of **non-additive measure are**:
- *Ratios* – e.g. ratios of men to women in different departments of a company
- *Measure of intensity* – e.g. temperature, speed
- *Percentages* – e.g. return on investment percent
- *Maximum* – e.g. temperature, blood pressure
- *Minimum* – e.g. income, account balance
- *Averages* – e.g. account balance, arrival time
- *Code* – [used for identification] e.g. social insurance number [SIN], postal code, barcode
- *Sequential numbers* – e.g. order number, ID number

Example of **semi-additive measure**:
- *Changing data* – e.g. address, telephone number
- *Dirty Data* – e.g. duplicated data, incorrect data

Averages, maximums, minimums and so on are a big problem, because they appear to be regular numbers but they are definitely not additive. Summing averages, or percentages, or ratios, or anything like that gives useless results that do not appear to be useless.

Let's say, for example, that we need to make a fact table with sales volumes reported by region, sales team, product, date, product supplier etc. We decide that we need daily totals and monthly totals, but we do not want to calculate the monthly numbers on the fly all the time, and since both will be using the same dimensions, why not put them in the same fact table. If we do this, then we can put an indicator in the fact table row, showing some records as daily and some as monthly, or we can just put the previous months' information duplicated for each day in an extra column in the daily rows. There are most likely other ways of doing this. The problem we have now is:

One, we need to remember to use "Select Distinct" to get the monthly data, which increases the complexity of the queries, and we need to make sure everyone who ever uses this table for as long as it is in existence knows about the requirement to use "Select Distinct".

Two, what if there is a problem? What if one of the days was reported wrong? Then we need to either update the rows in the fact table a [big no-no] or mark the rows as deleted and insert a bunch of new rows. Another problem – what if one of the dimensions changes part-way through the month? The daily data would still be valid, but the monthly totals would need to change part-way through.

The best solution to this problem would be to just use two fact tables, or even better, do the aggregation on the fly. These are just some of the issues that could be caused by mixed granularity.

In the past, applications were developed independently without any consideration that the data that were used in the applications would ever have to be integrated with other applications' data into the corporate data image. However, one important aspect of the Data Warehouse is the integration of all corporate data from multiple and very often incongruent data sources. As the data from online transaction processing [OLTP] systems, batch systems, and from externally syndicated data sources is fed into DW, it is being transformed and reformatted into one single corporate image for each entity. Because data from two different systems is to be merged together in the warehouse, it is obvious that we have to reconcile the different representations of the same data. This leads to an unintended benefit of an enterprise-wide process of data standardization and had been talked about for some time as a *desirable goal*, but no one has ever been given the responsibility of spearheading this effort. Drawing attention to these disparities is an eye-opener to many who have had no idea that other units of the enterprise were representing identical information quite differently. The diagram below illustrates this aspect of data integration as it passes from the operational environments or the external syndicated feeds into to the EDW.

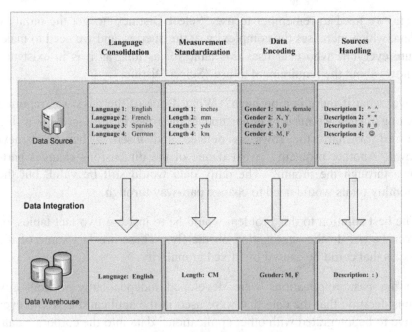

DATA INTEGRATION

"Data integration is the process of combining data residing at different sources and providing the user with a unified view of these data."

- Maurizio Lenzerini (2002). "Data Integration: A Theoretical Perspective". PODS 2002: 233-246.

Fig. 7.17 Data integration

- Table names should be intuitive, so a formal definition of the table may not be required but is nonetheless still useful.
- A table is best described by the data elements that reside within it. Each table should be identified by how it is to be used within the warehouse, e.g., fact table, dimension table or summary table.

- Primary keys should be identified and placed at the top of each table for quick reference.
- Validation that the primary keys in the dimension tables are actually unique identifies each row in a table
- It should be validated that the primary key in a dimension table uniquely identifies each row in the table.
- It is important that the primary keys of dimension tables remain stable. Therefore it is strongly recommended that the surrogate key, also known as an artificial or identity key, a substitution for the natural primary key, be created and used for primary keys for all dimension tables. The only requirement for a surrogate primary key is that it uniquely identifies each row in the table.

The surrogate key is useful because the natural primary key, e.g. the customer number in the customer table or the city name in the address table, may change. Source data tables may use different keys for the same entity. Legacy systems that are provided might have used a different numbering system than a current online transaction processing system. The surrogate key uniquely identifies each entity in the dimension table regardless of its source key. Slowly changing the dimension can only be implemented using a surrogate key. Dimension tables should be similar across all the fact tables.

Verification:
The second part of the test strategy should include testing or verification for:
1. *Sequence of jobs* - to be executed with job dependencies and scheduling
2. *Notifications of problems issued* – given to IT and generated in proper format
3. *Accuracy* - of the results.
4. *Re-startability of jobs* - in case of failure
5. *Generation of error logs and audit tables* - generated and populated properly
6. *Cleanup scripts* - for the environment including the database
7. *Granularity* - of data is as per design specifications

The combined responsibility for and participation in this activity by experts from all related applications is a must in order to avoid misinterpretations of the results.

When creating integration test scenarios, consider how the overall end-to-end process works and can break. Focus should be on intermediate points [points between the applications] between applications, not on any application involved. Testing should include consideration of process failures at each step, how the flexure is handled and how data would be recovered or deleted if necessary. The intermediate points [interfaces] are the areas where the defects are found.

Most defects found during integration testing are found, surprisingly, at interfaces.   Defects are either data related or resulting from false assumptions or misunderstandings about the design of another application.   Therefore, it is important to do the integration test with production-like data. Real production data is ideal, but depending on the content of the data, there could be privacy or security concerns that require certain fields to be randomized before using it in a test environment.   Communication between all groups, QA, BA and Development, is a must.   To bridge the communication gap, team members from all systems are asked to review and formulate test scenarios and discuss what could go wrong in production. There should be a dry run of the overall process from end to end in the same order and with the same dependencies as in production. Integration testing should be a combined effort and not solely the responsibility of the QA team.

Regression testing:
This is really, the equivalency testing, in which we re-run the same suite of tests to assure that the current version have not inadvertently regressed to the previous version especially in areas which to have been changed. During the test cycle, any time actual test results do not match the expected test results the program producing the error must be fixed, and all programs must be re-run.

A DW application is not a one-time solution. Possibly, it is the best example of an incremental design where requirements are enhanced and refined quite often based on business needs and feedbacks. In such a situation it is very critical to the test that the existing functionalities of a DW application are not messed up whenever an enhancement is made to it. Generally, this is done by running all functional tests for existing code whenever a new piece of code is introduced. However, a better strategy could be to preserve earlier test input data and result sets and run them again. At that point the new results could be compared against the older ones to ensure proper functionality.

Regression testing is a revalidation of existing functionality with each new version of code. Due to defect fixing and or enhancements in the DW application the regression test suite will be executed multiple times. That is why the regression test suite is a good candidate for automation, Start building regression/automation test suite during system testing. Regression testing process runs much smoother if the regression test suite has been automated. Test cases should be prioritized by risk in order to help determine which ones need to be re-run for each new release. A simple but effective and efficient strategy to retest basic functionality is to store source data sets and results from successful runs of the code and compare new test results with the previous runs. When doing a regression test, it is much quicker to compare results to a previous execution than to do entire data validation again. The recursion test, or any test that needs to be repeated, should be automate.

Performance testing:
In addition to the functional tests described above, a DW may need to go through another phase called performance testing. Any DW application must be designed to be scalable and robust. This must be verified before DW application goes to production environment. The performance concerns are in two primary areas:

1. The speed for data cleansing and staging after an accounting period has closed
2. Performance of data retrievals for the information consumers.

Performance test must be done with a large production-like volume of data. Objective is to ensure that the load window is met with production-like volumes. This phase should involve the DBA and ETL teams and others who may be able to review and validate the code for performance optimization.

The primary purpose of a DW application is to ensure that the data is accurate and credible. While this is necessary, it is not enough. Where DW application typically falls short is in delivering information value. Data is not information. The correct formula for information is relevance of data in that context. Providing information from the DW application into consumers' hands as they need it is a key part of providing information value. Data cleansing and staging may take a long time and by that time, the information may lose its value relevance.

As the volume of data in a Data Warehouse grows, the ETL load times can be expected to increase and performance of queries can be expected to degrade. Having a solid technical architecture and a good ETL design can mitigate this. The aim of the performance testing is to point out any potential weaknesses in the ETL design, such as reading a file multiple times or creating unnecessary intermediate files. The following testing scenarios would help discover performance issues:

- Load the database with peak expected production volumes to ensure that the ETL process within the agreed-upon window can load this volume of data.
- Compare these ETL loading times to loads performed with a smaller amount of data to anticipate scalability issues. Compare the ETL processing times component by component to point out any bottlenecks [areas of weakness].
- Monitor the timing of the reject process and consider how large volumes of rejected data would affect the performance.
- Perform simple and multiple join queries to validate query performance on large database volumes. Work with business users to develop sample queries and acceptable performance criteria for each query.

Materialized views in DW are built with performance improvement in mind. Unlike an ordinary view, which does not take up any storage space because they are generated on the fly, materialized views provide indirect access to table data by storing the results of an aggregation query in a separate schema object. A materialized view definition can include any number of aggregations: [SUM, COUNT(x), COUNT(*), COUNT(DISTINCT x), AVG, VARIANCE, STDDEV, MIN, and MAX]. Materialized summary tables use *aggregate keys* to define a hierarchy of aggregation. Several dimensions can be contained in each aggregate key.

Performance testing for retrievals at different level of aggregation is required to verify acceptable performance at all levels of AGREGATIONS?

Materialized views can be stored as summary tables in the same database as their base tables. This can improve query performance within OLTP systems. Most databases do not have enough CPU capacity to handle both OLTP and the heavy demands of OLAP processing. To ensure good performance, materialized views are made available in a separate schema or even on another machine than the OLTP system. The two databases are synchronized on a nightly [rather than real-time] basis.

User Acceptance Testing [UAT]:

"The map is not the territory".
"When the map and the territory disagree, believe the territory."
— Jerry Weinberg

Acceptance Testing is testing from the users' perspective, typically end-to-end, to verify the operability of every feature. UAT is done after the completed system is handed over from the developers to the customers or users. The purpose of acceptance testing is rather to give confidence that the system is working than to find errors. The reason d'être of a Data Warehouse application is to make data available to business users. Users know the data best, and their participation in the testing effort is essential to the success of a Data Warehouse. We build software for users. They're not [should not be] concerned with how the software works, how it's organized, or the cute programming tricks we used. They're only concerned with how it behaves and therefore, testing that behaviour must be central to this testing effort. Therefore, acceptance testing must be performed by business users. They are the most knowledgeable to validate the functionality of the Data Warehouse application. This also helps in avoiding "Yes, this is nice, but it would be nicer if..." syndrome. Quality Assurance assures and the development team support the UAT team during this test. This is the most critical part of the test cycle because the actual users are the best judges to ensure that the application works as expected by them. Business users, for example, do not have ETL knowledge. That is why they must be supported during this test by Quality Assurance and development teams. The UAT test exit report is signed off by the users and they must be able to understand it before they sign-off on it.

User-acceptance testing typically focuses on data loaded into the Data Warehouse and any views that have been created on top of tables.

1. Production [if possible] or near-production data should be used for UAT. Users typically think of issues in terms the "real" data.
2. Test database views are compared with expected results.
3. The QA test team must support users during UAT. This is because the end users will have questions about how the data is populated and need to understand details of how the ETL works.
4. During UAT, data will need to be loaded and refreshed a few times.

Risk Management

> *"The greatest risk in life is not taking any risks."*
> ~ Robin Sharma, author of the bestselling
> *The Monk Who Sold His Ferrari*

This is a true story that may have not happened, but ought to have happened, of how the greatest company in the world was shattered into mediocre existence by its own success.

Once upon a time, not so long ago, there was a great company great to its customers, great to cities and countries wherever it operated its business.

People all around the world admired this company. It was more than a company; it was an institution known for its passion to be right, its rigorous processes, its obsession for outguessing customers' desires, its thorough training of employees, and its "cradle-to-grave" employment policy. Yet the very same attributes that made the great company admired worldwide soon became its shortcomings. The proclaimed values that made the company famous could not sustain it a world where the marketplace was changing at the speed of internet. Soon the walls it had built to protect its employees became walls of imprisonment for those who remained within them for too long. The company became known for its aversion to risk-taking, its brainwashing of employees, its failure to invent new products and its inability or unwillingness to anticipate its customers' desires The company created a comfort zone for its professional employees that made them unwilling, unable or afraid to move out it. Occasionally some employees would snap out of the realm of their comfort zone, but people above and below would pull them back into their midst until eventually they succumbed to the rules of its civil service mentality.

We interviewed some ex-employees of this virtual company with the simple question:
        "Why did the great company collapse?"

The Answer was unanimous the upper echelons were to blame:
        "Senior management was fat and happy; the rank-and-file didn't have much say in the matter."
This statement taken at its face value is true. After all, noted authority Dr. Edward Deming stated convincingly that 96% of all the problems in

a company can be attributed directly to its management. The more interesting question is: **Why** didn't the leaders of the great imaginary company do their job? Presumably everyone wants to do a good job. The upper echelon is not an exception to this rule. The only logical explanation is: they did not know how!

Two developments back in 1960, more than anything else, shaped the future of the information technology industry as we know it today. First was the invention of microprocessors which led to the development of Personal Computers [PC] and the second was the development of Graphical User Interface [GUI].

Modern GUI was derived from PARC User Interface [PUI, also an acronym for **Perceptual** User Interface], developed by researchers at Xerox PARC [Palo Alto Research Center]. PARC user interface consisted of graphical elements such as windows, menus, radio buttons, check boxes and icons, and it also employed a pointing device in addition to a keyboard.

In 1979 Apple Computers started by Steve Jobs with some former members of the Xerox PARC group continued to develop the ideas of the Perceptual User Interface. The Macintosh, released in 1984, was the first commercially successful product to use a GUI desktop metaphor in which: files resembled pieces of paper; directories resembled file folders; and accessories appeared as calculators, notepads, and alarm clocks. The user could manipulate those metaphors around the screen as desired and could for example; delete files and folders by dragging them to a trash can on the screen.

In the meantime, our great imaginary company, even though it had been using its own microprocessors many years before they ware discovered by the rest of the industry, and even had a great pool of talented software engineers, who developed prototypes using this new technology, stood by doing nothing! Its upper management lacked vision - they just could not imagine the future with a PC on every desk and its impact on business and society in general. They could not see emerging patterns for new business opportunities in which every employee would be empowered by the technology they already possessed in their labs and would be able to utilizes their extensive knowledge and expertise of their employees. This coupled with their aversion for risk taking prevented them to seize upon an opportunity!

The processes of seeing patterns of the future, making new connections and risk taking to seize opportunities are basically creative right brain activities; and the more you try to structure and organize them, the more they disappear. Our imaginary company was run by left brain leadership which invested much energy into structure and organization, and owed its success to left brain thinking. As a result was unable to switch gears overnight reinventing itself into a new paradigm of whole brain thinking that was appearing behind the horizons.

In the collision between the creativity of the right brain and the logic of the left brain, subservient right brain thinking loses based on the final argument: "This is not the how the things are done around here." The more dominant left brain thinking prevails, by stamping over someone's dreams that ware like a delicate spring flower so innocently planted in mud before their feet. Eventually we learn the lesson of never planting flowers anywhere.

Continuing these causal analyses the same question remains: "But **why**?" It turns out that the problem runs much deeper into our society. Picasso once said that all children are born artists!!! When we send them to school, some are labeled as having learning disabilities. This is the story of a six-year old girl who was labeled as being unable to concentrate. In her drawing class, she was sitting in the last row, deeply concentrated in her drawing. The teacher came to her desk and asked, what she was drawing. Without taking her eyes of her work, she answers:

> *"I am drawing a picture of God".*
> *"How do you know, what God looks like? Nobody has seen Him?"*
> *"They will in a minute!"* She answered.

Later her parents explained it to her that if she wants to qualify for university and ultimately get a job, she must concentrate on math, science and English. This is how society eventually educates our children out of their natural creativity. By avoiding experimentation and risk taking, especially out of fear of what others may think, puts a limitation on our creative capability, as nothing new can be invented if one is not prepared to take a risk of making mistake. Our educational system is broken, and attempting to improve on something that is already broken is not going to correct the quintessence of the problem. We know the education system is broken otherwise we would not be continuously trying to improve it. We are even giving some fancy name to those programs, like "No child left behind", but sadly, the more things change, the more they look the same! Radical new thinking is required as Albert Einstein suggests:

> *We can't solve problems by using the same kind of thinking we used when we created them."*

The educational system narrowly focused on academic ability; inevitably marginalizes students with interests and gifts in other domains. As Socrates expressed it:

> *"Education is the kindling of a flame, not the filling of a vessel".*

Learning is a very personal process. Each student has different needs and learning styles. The purpose of teaching should be to ignite the students' imagination by supporting her or his passion and learning style. No one can be forced to learn against his or her free will. Yet, that is precisely what is being done currently. Students are compelled to learn outside their natural mental environment and ultimately pay the penalty of failure. For example, even the most reluctant students eventually learn to commit facts and ideas to memory, or drop out of the system.

Partly to blame for the adaptation of this faulty educational model is renowned Swiss developmental psychologist Jean Piaget who maintained that intuitive ability was prevalent only during the first two years of life, but afterward is overtaken by other powerful abstract and intellectual ways of "knowing". His "stage theory" of development had a profound impact on several generations of educators, who saw it as their job to redirect children's learning away from the reliance on their senses and their intuition while, at the same time, encouraging them to learn skills of deliberation and explanation sooner rather than later.

There is ample evidence that this kind of thinking is widespread in society and even more so in the corporate world [with some exceptions]. For example, a sophisticated mathematical model consisting of hundreds of partial differential equations of the national economy may be represented by a computer program which can measure everything that can be counted. Conversely anything that has no measure, like human nature, which is intangible [impossible to measure], has no value. In other words our assessments are like those of a man who was searching for his car keys under a streetlight, only because that is the only place where he could see.

Inherent in this mode of thinking is linguistic inadequacy of language used in describing business requirements, especially requirements for developing DW application. The fundamental rule of writing business requirements is that if something is to be understood it must be writhen clearly and unambiguously in order for the rational [logical] mind to understand and ultimately create a computer program. Language inadequacy poses a problem when it comes to describing evocative world of metaphors and imagery, most prominently exposed in describing DW application requirements. These kinds of problems require different types of "knowing". "Knowing" that emerges from not-knowing, uncertainty and mind's acceptance of this transitional state as those are the seeds from which creative thinking will sprout.

Western culture has lost the sense of this different ways of "knowing". These different ways of "knowing" are referred to as subconscious intelligence and conscious intelligence. Conscious intelligence is simply referred to as "intelligence", thus denying even the existence of any other intelligence. In making sense of the world, our common thinking process is dependant on pattern recognition based on our prior experiences. But to find something new we cannot depend on quick decision [rational thinking] as they are inhibitors to deeper intelligence [subconscious intelligence] since they are based on previous experience. Thinking creatively occurs when the mind slows and relaxes, thus activates a new and a unique pattern of neurons, forming novel association between them, only then the other ways of "knowing" automatically appears. William (Bill) Lear world famous American inventor and aeronautical designer, attributed for saying, that he had no education and that was why he had to use his brain, also said:

> *"One of the unfortunate things about our education system is that we do not teach students how to avail themselves of their subconscious capabilities."*

The New discipline of "cognitive science" association in the fields of neuroscience, philosophy, artificial intelligence and experimental psychology is confirming the existence of the subconscious intelligence of the human mind and that this subconscious intelligence is responsible for accomplishing the most unusual tasks [analyse and make sense of the most complex situation] providing it had enough time to do so. The most ingenious solution do not come about as a result of logical reasoning, they "occur to us". Complex mantel processes occur without our control or awareness. To the famous Italian opera composer Giacomo Pucini, while composing his opera *Madame Butterfly*, it appeared that:

"It was dictated to me by God; I was merely instrumental in putting it on paper and communication it to the public"

On the other hand if education was personalized to match each student's needs, fewer students would pull out of it. Some say that personalized education is an impossible pipe dream, as it would too expensive. But we argue that it would be more expensive not to. The fundamental point of reference must be that education is not a cost, but investment in our future. Based on the second premise that all students are curios by nature, it is clear that and that most efficient and profound learning takes place when it is initiated and pursued by the learner. All the students are creative if they are allowed to develop their unique talents within the appropriate educational environment and if we want to conceive future opinion leaders there must be a better ways to educate our children.

There is another more pragmatic reason for unleashing artistic right brain power within an organization. Artist those heralds of the future, create, not only for the present generation, but even for those that have not been born yet. A good artist is many years ahead of his or her time, or in the words of great Canadian hockey players, Wayne Gretzky: "I am where pack is going to be, not where it has been". "Artists' way of knowing" is not a special privilege of artists and sages; it is available and could be cultivated by the rest of us. Once when we where very yang, we new that and it is not that we have forgotten it, it is just that we don't trust it any longer.

The new business reality requires all of us to awaken the artistic power within organizations, which up till now has been suppressed and unwelcome. Leading MBA programs are introducing art and design in their curricula. Leader of companies must learn new skills if they want to conceive the new products to be ahead of the game, rather than waiting for their competitors' products to recognize the future.

If the above conclusion seems farfetched, then let's take a look at some examples of successful products in the marketplace today. The iPod, for example, was most certainly developed employing a minimalistic approach to design of the user interface. This example is evidence that the art and patterns of expansive thinking are nurtured in an organization where the collective aspiration is set free, as an integral part of the corporate culture. A similar example of this kind of the organization is Google. Word *Google* has become a new verb and not just in the English language. Companies that embrace art as the part of their corporate culture are prospering, while their closest competition is continuously playing a catch-up game.

In many situations our rational intelligence is necessary for our survival, yet there are always other ways to perceive and act then through our habitual way of thinking. Creative thinking is increasing freedom in our lives by providing other options and opportunities in problems resolution. Creativity is not about discovering new lands, but looking at the same land with a new set of eyes. It is about seeing the extraordinary within the ordinary. As the Tao Te Ching puts it:

> *"Truth waits for eyes unclouded by longing.*
> *Those who are bound by desire see only the outwards container."*

Risk is the possibility of suffering loss. Business must learn to anticipate and mitigate the possible negative consequences of risk against the potential benefits of its associated opportunity. In itself, risk is not bad; in fact, it is essential to progress. Failure is often a key part of learning too. Risk taking and the significance of learning from failure is often overlooked. Corporate culture that nurtures creativity is also tolerant to risk taking as it sets the stage for innovative thinking. Ingredients inherent to this culture that are, among other things, necessary for creativity, are the ability to challenge assumptions, recognize patterns, seeing out of the box, make new connections, take risks and seize the moment upon a chance.

The future is not the place we will go to, aimlessly like a ship with no rudder on the open ocean. The future is a place the leaders of a company must invent and clearly communicate and steer an organization in that direction.

Putting the blame on the educational system is convenient scapegoat for abolishing personal responsibility, as school should not be looked upon as the end, but the beginning of education. With or without the education each one of us has to make a choice, weather we want to live today as yesterday waiting for Fridays, or do we engage in adventurous journey into unknown full of possibilities, which eventually becomes that of what we think about. Thus proving that we are the masters of our destiny, as one thing that is always under our control is what is going on in our heads.

Strategy Review Time

> *"Person can have the greatest idea in the world –*
> *completely different and novel - bat if that person can't convince enough*
> *people, it doesn't matter."*
> ~ Gregory Berns[1]

QA Strategy document is an important document, but in this case, it differs from company's standard practises and procedures and it may be in grave danger if it is not it not effectively announced to the world. The stakeholders must be informed and convinced before they can commit to full support and funding required for the project to proceed. If there is no support for it, it will go into the waste basket, where many documents that cross our decks end up.

The message of the QA Strategy is too important to be wasted. But the Strategy document is hampered by one-dimensionality of written word. Socrates [Greek: Σωκράτης) one of the founders of Western philosophy did not write anything. He believed that writing is inferior form of communication in comparison to the power of oral argument. Written documents could not answer questions therefore could not stand up for themselves. The opportunity to present QA Strategy document is QA Strategy Review session, which is a required event in many organizations. Full advantage of this event must be taken, by effectively presenting the QA Strategy story. You have the podium and it's your show. Use it, but wait, not before you read the next section.

Power of storytelling

> *"People will forget what you said, people*
> *will forget what you did, but people will never forget how you made them*
> *feel."*
> ~ Maya Angelou[2]

Imagine human beings living in a underground cave, which has a mouth open towards the light and reaching all along the cave; here they have been from their childhood, and have their legs and necks chained so that they cannot move, and can only see before them, being prevented by the chains from turning round their heads. Above and behind them a fire is

---

[1] Gregory S. Berns is a distinguished American neuroscientist professor of psychiatry and writer

[2] Marguerite Ann Johnson is an author and poet who has been called "America's most visible

blazing at a distance, and between the fire and the prisoners there is a raised way; and you will see, if you look, a low wall built along the way, like the screen which marionette players have in front of them, over which they show the puppets. The wall conceals the puppeteers while they manipulate their puppets above it. Imagine further man behind the wall carrying all sorts of objects along its length, and holding them above it. The objects include human and animal images made of stone and wood and all other material.

Imagine now one of the prisoners is freed from shackles. He turns around and walks towards the lights. He suffers pain and distress from the glare of the lights. So dazzled is he that he cannot even discern the very objects whose shadows he used to be able to see. He is blinded by light, actually by the truth, but unable to comprehend it.

He is dragged away by some force up the steep incline of the cave passageway, until he hauled out into the light of the sun.

He goes back to the cave, where the others are shadow watching to reveal the truth to them. The brave sole who risked all to seek the truth is now being ridiculed for even attempting the journey!

This of course is the famous *"Allegory of the Cave"* extracted form Plato's *"The Republic"*. "The *Allegory of the Cave"* represents an extended metaphor for the state of human existence, and for the transformation that occurs during enlightenment. By using storytelling Plato hopes to enables the reader to embrace the full potential that has been living beneath the surface. When the light of the sun shines on the freed man is the metaphor for enlightenment and perception of the truth. In "The Allegory of the Cave", Plato brings to light the idea that without education, in this case through stories, people are like prisoners in a cave living amidst shadows. People without knowledge know nothing about experience or desire. Their reality consists of the images that dance before them on the walls of their cave. Yet this cave is not a physical one; this cave is an imprisonment of the mind. To Plato, the most important thing in life is to be educated continuously in the hopes of better understanding life and all that is a part of it.

Upon returning to the Cave, our escaped prisoner would metaphorically be entering a world of darkness again, and would be faced with the other chained prisoners. The other prisoners would laugh at the released prisoner, and ridicule him for even taking the useless ascent out of the cave in the first place. The others cannot understand something they have not experienced, so it's up to this prisoner to epitomize leadership, for it is him alone who is conscious of goodness. It's at this point that Plato describes the philosopher kings who have seen the Ultimate Goodness as having a sense of duty to be responsible leaders and to not feel disdain for those whom don't share his enlightenment.

Most people, including ourselves, live in a world of relative ignorance. We are even comfortable with that ignorance, because it is the best we know. When we first start facing truth, the process may be frightening, and many people run back to their old ways. But if you continue to seek truth, you will eventually be able to handle it better. In fact, you want more of it! It's true that many people around you now may think you are weird or even a danger to society, but you don't care. Once you've tasted the truth, you won't ever want to go back to being ignorant!

The aspect of the above story we are interested in here is: Now that you have seen the "light" how do you move other out of their "comfort zone" and how do you confront all those "I-told-you-so-ers" and "naysayer", who will offer many reasons why that won't work and who will kill you at the end if you don't convince otherwise.

The prescription for the solution to this dilemma came from Plato's student, Aristotle. Plato devised and used his "Allegory of the Cave" as a teaching tool and Aristotle, who ought to have been part of those discussion, turned to the most ancient art of storytelling, to resolve the dilemma, of how to retune to the cave and convince the fellow cavedwellers, that at the end of that long passage into their cave there is the most beautiful world – a promised land. Aristotle in his book "The Poetics" has reviled the "secret" of story-telling to us, more than twenty three centuries ago. The story has to appeal to emotions [pathos], reason [or logical appeal – logos] and personal credibility [ethical appeal – ethos]. Facts alone are not sufficient to persuade. The most effective way to convey the most complex idea is to turn it to a story.

Sufi mystics have for centuries instructed their followers with stories. Many Sufi tales have been passed down the centuries into legends and ethical teachings. The tale of the elephants used earlier in "Story of Quality Assurance" section of this book is a famous example. Islamic tradition of Sufi tales entertains while teaches at the same time. The goal of stories is to amplify perception and helping to gain deeper knowledge. Only a few Sufi stories can be read by anyone and still unlock internal consciousness. Sufi masters unlock the internal dimension of the stories when he feels their disciples are ready.

Stories have been used since time immemorial for communicating important messages, to inspiring, or sharing knowledge and ideas. Everyone responds to a good story. For as long as humans have existed it has been the way of relating, history, legends and lessons, sometimes to entertain, or even to save lives as is an old Persian story of Scheherazade (Farsi: داستانهای هزارو یک شب شهرزاد قصه گو). This ought to be a true story as she lived to tell it to us:

Scheherazade was a Persian queen and the narrator of One *Thousand and One Nights* [also called: *"The Arabian Nights"*]. The legends of the Arabian Nights were passed down through the centuries by word of mouth; the oldest tales date to the 10th century.

It is about a Sultan who upon discovery of his wife in orgy with her slaves vows to never trust women again. To protect himself from the future pain he was in the habit of marring a virgin and killing his wives after the first night. This was going on for three years, until Scheherazade conceived a plan to stop him and convinced her father to offer her as the sultan's next wife.

Scheherazade was the oldest daughter of the grand vizier to Sultan Shahryar. Scheherazade was very well-educated young lady, having studied the legends, books, histories, and stories about Kings and civilization in general. She had learned all about philosophy, poetry and arts. Not only was she well read, but also was well bred, polite, and pleasant to all she encountered.

The night after their wedding ceremony, Scheherazade told Sultan her first story. Sultan Shahryar listened in awe as Scheherazade spun a fascinating, adventurous tale, but she stopped speaking before the story was finished. The Sultan asked her to finish the tale, but Scheherazade said there was no time left because it was almost dawn and time for her beheading. She added that she really regretted not finishing this story, because her next story was even more thrilling.

Sultan Shahryar decided not to execute Scheherazade that morning just so he could hear the rest of the story later that night. Scheherazade began one of the exciting tales but stopped before the story ended, causing the sultan, who had listened as well, to put off killing her until she could finish her story the next evening. Scheherazade, of course, never finished her tales, but kept her husband enthralled with story after story for 1,001 nights.

Scheherazade must have known that here storytelling was not to show up how brilliant she was, but about making Sultan the hero. While presenting her stories she was humble almost invisible, as her success depended on Sultan not the other way around. At the end of one thousand and one nights and one thousand stories, she knew she has reached remarkable outcome. Scheherazade told the Sultan that she had no more tales to tell him. During these one thousand and one nights, the Sultan had fallen in love with Scheherazade, and had three sons with her. Having been made a wiser and kinder man by Scheherazade and her tales, he not only spared her life, but made her his Queen.

Storytelling in Business

> *"This report, by its very length, defends itself against the risk of being read.*
> ~ Winston Churchill

Despite of what we have been told, the greatest invention of all time isn't the wheel, it's business organization. Communication is life blood of a business organization. Effective communication is essential for the prosperity and survival of a business. It is also a basic tool for motivation and education of the employees in an organization. The power of storytelling is such a potent communication tool, that it makes the need, to learn how to do it better, a must for business professionals. Every great story is about problems and problem resolution because that is vary much what the life is all about. The story of QA Strategy is not different; it presents the problem and its resolution.

In business we are exposed to a variety of stories on a daily basis, yet we do not fully appreciate the power of "storytelling" in reframing people's viewpoints. How many times at business meetings have one or more participants insisted that each had the correct perception of the issue at hand? Arguments are prolonged as they try to defend their respective positions, unwilling to entertain the possibility of an alternative point of view or to see the problems through a different lens. There is also the resistance to change, the resistance to seek new methods to resolve business problems because participants are often caught up in the message of the old stories which have worked so well in the past. It is very much like the diagram of the cube below. Once we see it from one perspective, it is difficult to see it from another perception.

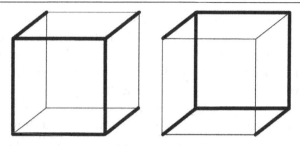

Fig. 8.1 Cubes

PowerPoint presentations have become the norm in business today. During a typical PowerPoint presentation the speaker stands before an audience and reads out the presentation – bullet point after bullet point. This style of presentation is the main reason why many business presentations are dismal failures – they also out of sync with how people communicate and learn. Edward Tufte, the visual communication guru and author of wonderful books like *"Beautiful Evidence"* and *"Visual Explanation"*, puts the blame directly on PowerPoint. In the September 2003 issue of Wired Magazine in the article entitled: *"PowerPoint is Evil"*, Tufte wrote:

> *"a presentation format should do no harm. Yet PowerPoint style routinely disrupts, dominates, and trivialises the content . Thus PowerPoint presentations too often resemble a school play - very loud, very slow, and very simple"*

Dr. John Sweller of the University of South Wales, who developed the cognitive load theory[3], discovered that it is more difficult to process information if it is simultaneously both verbal and visual. In other words, people cannot listen and read at the same time. The assumption made is that the PowerPoint presentation which includes lines of text projected on a screen mirrors the spoken words. As a result, reading PowerPoint slides to an audience can be ineffective in communicating a message. Poorly designed presentations do not achieve the presenter's intention and are not an enjoyable or effective learning experience for the audience.

---

[3] The term **cognitive load** is used in cognitive psychology to illustrate how the load is related to the executive control of working memory. People learn better when they can build on what they already understand. The more a person has to learn in a shorter period, the more difficult it is to process that information in working memory

Making a presentation should not be viewed as an opportunity to for the presenter to demonstrate his "brilliance". Rather it is an opportunity to make a small contribution or to share an idea that that the presenter feels is important. Millions of mediocre PowerPoint presentations are delivered every day, but the good news is that the bar is now so low that there is plenty of opportunity to make a difference with even a small improvement! A presentation that goes well will have a strong impact on the presenter's spirit, confidence and career. A good presentation means the "story" had been told effectively.

> *"The single most important thing you can do dramatically improve your presentation is to have a story to tell before you work on your PowerPoint file"*
> Cliff Atkinson[4]

Despite the overwhelming evidence that presentations using PowerPoint [or Keynote] should be abandoned, we will prove, by method of exhibiting an example, that PowerPoint presentations can still be effective. Steve Jobs' *"2007 iPhone launch*[5]*"* presentation is an excellent example! It is a clear, visually supported story in which today's world and his vision of an improved future world are contrasted. His presentation consists of slides without bullet-points. Slides contain only pictures to support Jobs' story.

It is the "story" and the metaphors used in the story that convey the message, bullet-points on a PowerPoint [Keynote in this case] do not engage conceptual thinking. Why is this? The audience responds best to tangible, moveable and inspirational things. Humans make decisions based on emotion, and then look for the facts to support their decisions or choices. Facts alone just can't do that.

This is supported with the recommendations of the latest findings in cognitive theory. The research of Dr. Richard Mayer, educational psychologist at the University of California, which has been reported in a study named *"A Cognitive Theory of Multimedia Learning"* in which he outlines the fundamental principles of multimedia design based on cognitive functioning. He writes:

---

[4] Cliff Atkinson, *Beyond Bullitt Points*
[5] Steve Jobs announcing the new iPhone. Apple Reinvents The Phone

"It is better to present an explanation in words and pictures than solely in words...
When giving multimedia presentations present corresponding words and pictures contiguously rather than separately.... When giving multimedia explanations use fewer rather than many extraneous words and pictures."

2-PAGE SHORT STORY in TWO DAYS?

Mark Twain, prolific American writer, received this telegram from his publisher:

NEED 2-PAGE SHORT STORY TWO DAYS? Twain replied:
NO CAN DO 2 PAGES TWO DAYS. CAN DO 30 PAGES 2 DAYS.
NEED 30 DAYS TO DO 2 PAGES.

A shorter presentation with relevant information is more in tune with cognitive learning theories. Interestingly, this is also a fundamental Zen concept known as "kanso" or simplicity. This reminds us of the Japanese minimalistic approach discussed earlier in this book. Japanese Zen arts claim that it is possible to exhibit great beauty and convey powerful messages through "kanso". [Jobs' message of "a thousand songs in a pocket" sounds very much like a "kanso" message.] This is the same principle supported by Dr. Edward Tufte in eliminating "chartjunk" - eliminating superfluous design elements. On the same topic German painter Hans Hofmann said:

"The ability to simplify means to eliminate the unnecessary so that the necessary may speak."

The "minimalistic aesthetic" approach is evident in all Apple products. Time and time again, we see the same approach in Steve Jobs' presentations.

"One of the most important parts of Apple's design process is simplification"
Leander Kahney[6]

---

[6] Leander Kahney, *Inside Steve's Brain*

Indeed, the art of storytelling is an art form which, more than any other art form is integral to the human experience. When presentations are delivered in a story framework they put on view in contest of the subject of the delivery making it a formidable motivator for the audience. However, few of us have the public-speaking confidence of Steve Jobs. Knowing the rules of the art does not necessarily make an artist a "presenter". It should not come as a surprise that confidence is the result of hours of relentless practice. In fact, confidence is so important to the quality and effectiveness of a presentation that US president Barack Obama once said that the most important lesson he learned from working with the most powerful person on the planet was to "always act confident". Ultimately, the result of countless hours of rehearsal in any field of endeavour is confidence and effortless performance.

"Storytelling" is an art that can be learned. Even the world's geniuses, such as Wolfgang Amadeus Mozart, who composed from the age of five and performed before European royalty, had to invest hundreds of hours to master his art form. In his book *"Genius Explained"* psychologist Michael Howe[7] explains that to be a genius demands a strong sense of direction and an extraordinary degree of commitment, focus, practice, gruelling training and drive

> *"... by standards of a mature composer, Mozart's early works are not outstanding. The earliest pieces were all probably written down by his father, and perhaps improved in the process. Many of Wolfgang's childhood compositions, such as the first seven of his concertos for piano and orchestra, are largely arrangements of works by other composers. Of those concertos that only contain music original to Mozart, the earliest is now regarded as a masterwork [No-9, K.271] was not composed until he was twenty-one: by that time Mozart had already been composing concertos for ten years."*

If Mozart had not persisted as long as he did, the world would have never known the beautiful arias from "The Marriage of Figaro" or "The Magic Flute". We continue to enjoy y the benefits of his work today, according to a landmark neuroscience research study out of University of California[8] listening to one of the most profound and most mature

---

[7] Michael J.A. Howe is professor of psychology at the University of Exeter

[8] Center for the Neurobiology of Learning and Memory, University of California, Irvine,

Mozart's composition "Sonata for two pianos Concerto, K. 488" produces significant short-term enhancement of spatial-temporal reasoning in college students. Persistence and perseverance require passion and continuous encouragement for achieving mastery in one's chosen calling. For that reason it is important for each of us to discover our true passion as it is an absolute prerequisite for stretching our imagination and reclaiming our brain creative power and putting in service to whatever we value most and in that process realize our full potential.

First, we need to search for what is truly beautiful, whether it is a perfect piece of poetry, or piece of music, or laughing with a loved one and experiencing the deepest feelings of empathy. We need to associate with good people who tell good stories about other good people because the people we surround ourselves with have a profound effect on who we are and who we become. In other words, we need to learn to relate to other human beings in a positive way. A beautifully told story is a symphonic unity in which structure, setting, character, genre and where ideas melt seamlessly. In order to find their harmony, the author must study the elements of the story, as if they were instruments of an orchestra - first separately, then in concert. Life is too short to be wasted in associating with people who dismiss our passion and discourage our ideas.

Second, the relentless practice for many hours over many days is required to pursue and eventually to master our life's calling. Neuroscientist Daniel Levitin claims that the magic number in achieving mastery in any task is ten thousand hours:

> *"The emerging picture of such studies is that ten thousand hours of practice is required to achieve the level of mastery associated with being a world-class expert in anything... In study after study of composers, basketball players, fiction writers, pianists, chess players, master criminals, and what have you, this number comes up again and again. Of course this does not explain why some people don't seem to get anywhere with their practice and why some people seem to get more out of their practice session than others, but no one has yet found a case in which true world-class expertise was accomplished in less time. It seems that it*

Department of Physics, University of California, Irvine, CA 92717, USA 18 October 1994

takes the brain this long to assimilate all that it needs to know to achieve true mastery."[9]

To put this period 10,000 hours in prospective, it is equivalent to 10 years of performing the task repeatedly three hours per day. We now have the scientific proof for what we have known for years, and indeed as old adage goes: practice make perfect [Lt. Repetitio est mater studiorum]. Rehearse and re-rehearse! This is not different with skill of communication, as the presentation can not be better than the preparation effort that went into it.

In his book *"Brain Rules"* John Medina[10] [under rule #5 - "Repeat to remember"[11]] claims that it is that period, of approximately ten years of repeating a task, that is required for transferring temporary memories into more persistent forms. Medina told the story of an experiment in which a patient's hippocampus was removed because of a life threatening disease. After the operation the patient could not recollect anything from the last ten years of his life, but he had a clear memory of events from eleven years and earlier prior to the removal of the hippocampus. This proves that the patient's temporary memory, not yet committed to his more persistent forms of memory, was completely erased by removing patient's hippocampus.

Story-telling language is metaphor as that is the script language that directly communicates with the conceptual mind. The essence of metaphor is to understand one unfamiliar concept by relating it to the familiar one. Learning of new concept is enabled by quickly associating it to a familiar knowledge. The new idea is experienced by relating it to a familiar one. It's the conceptual system that defines the reality. All communication is interpreted by the same conceptual system which is also used in thinking and acting. Therefore if we want our audience to quickly grasp the contents of our story we must use metaphors.

The presenter's objective is to fully engage brains of the audience [stakeholders]. Before making a presentation presenter should know something about the audience. Their thinking styles, extraverts,

---

[9] Daniel Levitin: *This is Your Brain on Music*
[10] John Medina: Brain Rules: 12 Principles for Surviving and Thriving at Work, Home, and School
[11] From the book *"Brain Rules"*- Rule #5: Repeat to remember.

introverts, detail-oriented, conceptual types etc. More importantly is to know what makes them tick, in order to connect with them. The presenter's goal is to unify the group of individual who have temporary come together to hear presenter's story. In order to make the audience focus to the presentation, presenter must learn something about them, there skills, a few names, there weaknesses, demographics, etc. as this is where the empathy for the audience will come from. Otherwise as Ken Haemer of ATT says:

> *"Designing a presentation without an audience in mind is like writing a love letter and addressing it: To Whom It May Concern."*
> – Ken Haemer

Based on the above discussion, an effective presentation must still follow the story line prescribed by Aristotle many centuries ago: a well-designed story must have a beginning, a middle and an ending. The beginning sets the stage for the story where the emotional contact is made with the audience and where the problem is introduced. The middle part of the story develops the action where the main protagonist [hero] is faced with obstacles which prevent him or her from achieving a desired outcome. The ending is reserved for the resolution of the problem.

Every presentation requires a blueprint - a destination on a road map to show how to reach the desired outcome, the place where the audience's understanding and commitment are intended to be following the presentation. If the audience does not understand the presenter's blueprint, they will soon disconnect from the conclusion of presentation, both mentally and emotionally. The presenter's goal is to "transport" the audience from one location to another, or to persuade them to let go of the old ways and adapt new ones. Two questions must be answered for the audience if the presenter wants win their attention: "What is your point?" and "What's in it for us?"

The beginning is the critical part of the story. This is where Aristotle's classic power of establishment is engaged appealing to emotions and reason based on personal integrity. This is where the protagonists and their relationships are described and the "hero quest" is established. The intent of the first slide of the presentation is to immediately draw in the audience and focus their attention on the presentation. The questions to answer at this point are: "Where are we now?" and "Who are we in this setting?"

After making an emotional connection and presenting the problem clearly in the first part of the presentation, the audience is ready to hear the appeal to reason, in the mid part of the presentation. They are at this point closely paying attention to learn how **they** can solve the problem. The how and the why provide justification for their decision to accept the proposed solution in the first part of the presentation and at the same time this is the reason why they have come to here the presentation.

Ending part of the presentation is where everything, emotional and rational is tie together. Stage is set up for the audience - the protagonists of the story to hear answer to a question they have at this point: "What needs to happen so that we can resolve the situation?"

Journey through Bermuda Triangle

Visual analogy with the ships and voyage through the Bermuda Triangle is the subject of the conceptual metaphor in the visual story.

To set up the scene the title of this slide is: **"The ships are safe in the harbour, but that is not what the ships are built for."**

Visual: The body of the slide of a sailing ship docked in a tranquil harbor.

Narrative[12]: All organizations go through a life cycle of starting up, growing, maturing and eventually declining unless, of course, they "reinvent" themselves. [This is known as a concept of S-curve: new businesses start out slow, and then scale rapidly, the taper off]. Successful businesses are founded on the idea of the world, and the future world as an improved place. It requires courage and communication to recognize the declining cycle long before it is obvious to others. It also takes courage to move foreword to an unknown future which can yield both risks and rewards. Regardless of the uncertainty of the future, businesses must move forward in order to survive. Companies must learn to live and prosper in the chronic tension of what is and what could be. We are in the middle of the paradigm shift. This is not dissimilar to the stories of Bill Gates and Steve Jobs. Both of them

---

[12] The narrative text is entered in the notes part of the PowerPoint slide. If you must read your notes, Apple's presentation software, Keynote, allows speakers to see the notes, while the audience sees just the slides displayed on the projector screen. But it is more important maintaining the eye contact with the audience to reading the notes. Research has shown that eye contact is associated with integrity and trustworthiness. Conversely avoiding eye contact is associated with lack of confidence and sincerity.

toiled away in relatively obscure field of microcomputer without any hopes for any worldly success. But then it happened – the personal computer paradigm shift! They were ready; they had their 10.000 hours in. The world changed and they led the change, while the major computer manufactures stood by wandering what happened.

The second slide should introduce the "hero". At this point the audience will ask: "Who are we in this scene?" The presenter is not the hero; the presenter is there to support the audience and to help them recognize that they are the "real heroes" of the story. This slide also introduces the adversity and as Author, Robert McKee in his book *"Story[13]"* writes:

> *"Something must be at stake that convinces the audience that a great deal will be lost if the hero does not obtain his goal".*

So here it is, title of the second slide which identify the hero and the adversity is: **"Future of your company is in your hands"**

Visual: will have diagram of S-curve upon the S-curve, showing the paradigm shits.

Narrative: We today again have a choice, to maintaining the status quo by pretending that nothing is happening [and disappear into oblivion], or embrace the new BI reality. Inaction is not a smart choice, as we will soon find ourselves on steep declining curve, wandering what happened to our organization and consequently to every one of us.

After the paradigm shift trying to climb the old S-curve will no longer work. We must make the jump from our current S-curve to a future S-curve now. If our intention is to compete in the future marketplace and prosper in it, perusing BI has no alternative. Quest for BI is long and never ending journey that we all have to make commitment to if we intend to find a new prosperity in the new marketplace on the personal level too!

The next slide title introduces the imbalance with the title: **"BI today is sailing through the Bermuda Triangle"**

Visual: Ship zigzagging through the stormy sees towards its destination.

---

[13] Robert McKee *Story: Substance, Structure, Style and The Principles of Screenwriting*

Narrative: But journey to prosperity leads trough some dangerous waters. In fact so dangerous that nine out of ten ships are lost forever." The problem is caused by proliferation of one size fit all solution.

The cause of the problem is introduced in the next slide which should answer the question: "What stands between us and our dreams?" This also is the title in the heading in the next slide: **"What stands between us and our dreams?"**

Visual: Picture of summer sky in the evening with billions upon billions of stars...

The presenter's narrative begins: In summer evenings humans have always looked up and wondered about the mystery of the billions upon billions of the stars in the sky about which we know so little. There are also that many billions of neurons flying around the most sophisticated machine in the universe, inside our brains. Yet we know very little about our brains. However, one thing we do know for certain is that intelligence is not "out there"; it is inside our brains just waiting to be unleashed! Business or any other intelligence is simply not "out there", rather it has to occurs inside that most sophisticated machine - our brains. And that is exactly what we want to do – by unleashing human intelligence achieve BI. There is nobody else from outside who understands our business better than we do and who can do it better then we can do! The proliferation of the one size-fits-all solution is the cause of the problem and that by the way is the reason why Business Intelligence today is contradiction in terms [an oxymoron]!

This will follow the slide stating the desired outcome: **"In pursuit of wisdom – journey without ending."**

The body of the slide [visual] reflects a zigzag path of the ship from the start point to the destination, and additional path which the ship would have taken without onboard navigational system. This metaphorically relates the value of the prototype to an onboard navigational system. In other words, an onboard navigational system is enabling the crew to avoid the obstacles along the way, but at the same time, know where we are all along the path in relation to the desired destination.

Narrative: The secret is that the intelligence and wisdom to create BI for your organization is inside every one of you. Using previously discussed Excel prototype, base on something familiar, we will be able build something new - BI for your organization. The awareness of knowing our location is continuously obtained from the business user, thus forming the coherent team of business and IT experts moving and learning together through an uncharted sea. In the process of moving and learning together human intelligence is unleashed, a process that can never be even initiated by any third party software.

The learning process will be enhanced and eventually mastered by many repetitions of this cycle as we repeatedly navigate through the stormy seas of the business intelligence, similar to the repeated rehearsals of the masters in the cycle of 10,000 hours. The ending of one cycle is simultaneously the beginning of the next cycle of the journey for the BI team. And so it is repeated ad infinitum.

> *"What we call the beginning is often the end.*
> *And to make an end is to make a beginning.*
> *The end is where we started from"*
> T. S. Eliot

It is much later than you think!
During the research for his book "Inside Steve's Brain", Leander Kahney was struck by Jobs' apparent preoccupation with death, signified by how many times he mentioned it. This may be the driving force behind Jobs' urgency to "dent the universe". In a commencement speech to the graduating class at Stanford in 2005, he observed:

> *"Remembering that you are going to die is the best way I know to avoid the trap of thinking you have something to lose. You are already naked. There is no reason not to follow your heart"*

In the book you are reading, it's not by accident that the terms used throughout it like, *battlefield, warriors, strategy, trenches, echelon, etc,* are all metaphors with one common denominator: The conceptual subject of those metaphors is war. Those metaphors *are* used in this book because they convey the urgency and importance of the subject and its consequences! It's just that the process is too gradual and hard to be noticed, but the evidence is convincing: we are loosing 10% per year of well paid "white collar" jobs every year, our houses are being repossessed at an alarming rate, soon we will not be able to afford our educational and health systems that we have benefited so fur! We are facing "silent crisis", in which things are changing in real and meaningful ways, but such changes are hard to detect in our everyday lives on a day-by-day basis. Maybe you've heard of the experiment with a frog before, but let's use it here to make this point clearer:

> *Throw a frog in a pot of water. If you turn the heat up on the water quickly, the frog will sense the temperature change and jump out of the pot. But if you raise the temperature slowly, the frog won't notice until it's cooked, but then it's too late to do anything about it.*

Pretending nothing is happening is not an astute strategy. First we must find out what the cause [not the consequences, as those are the just the symptoms] of the alleged problems described in the above paragraph is. The author Thomas Friedman seams to know the answer. In his best-selling book"*The World Is Flat[14]*", he attributes this to a phenomenon which he calls *globalization*. Friedman, while on a journey to India, realized that *globalization* has changed core economic landscape of the world. In his opinion, this "flattening of the world" is a result combination of number of coincidental events like, crumbling of borders [collapse of the Berlin Wall], convergence of the advances in fiber-optic networking coupled with personal computer, etc. Friedman's "Flat World" is probably his metaphor which relates to the work, which is becoming commoditized and sold in global markets. He should [but he does not] further refer to the commoditization of only the mundane work in manufacturing and services, or reading manuals at call centers over the phone, etc, but excluding not-so-mundane intellectual activities, like product innovations and high-paying non-contestable creative jobs like those required in development of Business Intelligence.

It is neither globalization nor a "flat world" that is causing problems in our labour market [but it's a nice scapegoat]. It is the paradigm shift! Developed countries are moving away from the "economics of goods" to the "economics of information". It's technology and the new requirements of post-industrial society that is causing labour market shift. While it is true that many skills required for prospering in the industrial age are no longer required. But at the same time this paradigm shift has created tremendous opportunities for those that ware ready for it. Adaptation of deep new knowledge is required and the need to learn fast in order to stay relevant and prosper in the new labour market. It appears that paradigm shift rules apply equally well to everyone of as they do to organizations. Instead of blaming our misfortune to circumstances beyond our control, it's better to embarrass the new reality

---

[14] *The World Is Flat: A Brief History of the Twenty-First Century* is an international bestselling book by Thomas Friedman

as the old one is gone forever. The proof of this premise is that many jobs go unfilled due to lack of qualified professionals, this seam to coincide with the opinion of qualified and respectable source in their "MCKINSEY QUARTERLY MONTHLY NEWSLETTER JULY 2011[15]":

> **Big data: The next frontier for competition**
> *The ability to work with vast data sets--big data--could spur productivity and innovation, but the United States alone has a shortage of up to 190,000 people with suitable training in statistics and machine learning. Our interactive exhibit, on the McKinsey & Company Web site, explores the industries and occupational categories where these specialists work.*

On a final note, the reader should remember that these guidelines are only points to ponder and not mandates. Art can open established doctrines available not only to artists but engineers and programmers and it should be used by them when the circumstances dictate. Effective PowerPoint presentations exist as a new type of art, form heralded by interested presenters, and certain rules can and should be broken if the situation warrants a change.

"**Just one last thing**" [Jobs] remember the 10,000 hours rule, the sooner we start packing in those hours, the sooner we will become experts in BI.

---

[15] http://www.mckinsey.com/en/Features/Big_Data.aspx

Testing - Moving from "What" to "How"

> *"In order to find bugs, we have to believe in the possibility that*
> *they're there"*
> ~ Glenford Myers

Testing is a skill and it is extremely creative and intellectually challenging work. While this may come as a surprise to many people, this is a simple axiom. The purpose of testing is to run a set of test cases against the program under test, with the sole goal of finding defects. Establishing the psychological goal of finding defects implies the tester's assumption that there are defects in the application. The other necessary assumption for success in testing is the notion of "successful" and "unsuccessful" tests. From the testing perspective, [project managers may have a different view] any test case that uncovers a defect is a successful one; any other test is not. This is no different in the practice of medicine: a test run by a medical doctor that discovers an illness is considered positive, and those that don't are labelled by the medical profession as negative results[1].

For any moderately complex software application, there are an astronomically large number of possible test cases, yet at the same time, there is time to execute only a fraction of them. Thus, a small subset of the all possible test cases is expected to find most of the defects in the application. This requires building good test cases, test cases with a high probability of finding defects.

Attributes of good test cases

Besides the input conditions, a test case must also include the expected results. Testing without knowing the expected results is not testing but experimenting. A test set must also include test cases with invalid conditions [e.g. the numeric field may be tested with alpha or special characters] to verify that the program being tested is not doing something that it is not supposed to do [hard crash, or program goes to land of no return]. For example, a transformation program may be reversing a list of numbers before loading them. As a part of testing, empty list is passed to program and the program blows up. The requirements don't stipulate that program have to accept an empty list. Program, must never blow up, at

---

[1] Glenford Myers: *The Art of Software Testing*

very minimum should throw an exception if the routine is called with an empty list.

A test case with all input fields invalid is not a good test case as it is hard to tell which input created which response. A test set must include test cases for interface and boundary condition testing, as that is where the defects are found. A good test case should test more than one function, thus reducing the number of test cases to build and execute tests thus increasing the probability of finding defects that may be functionally interrelated.

Good places to start finding test cases are business requirements, activity diagrams, use cases and design specification. This knowledge not only helps in finding critical test cases, e.g. at the boundary conditions and program interfaces, but it also helps in finding and eliminating equivalency test cases [if a test case tests the same feature or finds the same defect as does one other test case, it is called an equivalency test case and it can be eliminated]. A thoughtful approach is required in selecting and running the set of test cases without compromising the quality of the application.

A good rule of thumb for finding more defects in any program is that the probability of finding more defects in a section of code is directly proportional to the number of defects already found in that section of code.

The skill of testing is to find the maximum number of defects with the sufficient number of effectively designed test cases and at the same time, to keep the cost of testing at an acceptable level.
The final question is: when to stop testing? How do we know we have done enough testing?

1. End of schedule?
2. Rate of finding defects has flattened?
3. Management says: "Stop it!" - No more money in budget for testing?
4. All test cases have been executed and all critical defects fixed?

Exit criteria should be part of any test plan, but things don't always happen according to plan. In his book *"The Complete Guide to Software Testing*[2]*"*, Bill Hetzel writes regarding criteria for stopping a system test:

---

[2] Bill Hetzel: *The Complete Guide to Software Testing*

"Testing ends when we have measured system capabilities and corrected enough of the problems to have confidence that we are ready to run the acceptance test."

Boris Beizer's stopping criteria is even more prone to interpretation:

"There is no single, valid, rational criterion for stopping. Furthermore, given any set of applicable criteria, how exactly each is weighted depends very much upon the product, the environment, the culture and the attitude to risk."

More practical test stopping criteria is based on the notion of test coverage. For example, project stopping criteria [depending on the project] could be defined as:

- 100% coverage of use case scenario, and/or
- 90% coverage of branch coverage scenarios.

At the level of integration testing, stopping criteria may be defended in terms of API testing coverage.

According to Glenford Myers, the coverage-based stopping criteria is "highly counterproductive", as it does not motivate testers to create a strong test case that would find the defects. In fact, it encourages them to write test cases with a low probability of finding defects, in order to meet the coverage criteria.

The Case for Automated Testing

Test automation is a critical regression activity for iterative development where functionality builds up from iteration to iteration and releases cycles that are shorter, but at the same time, the size of the regression tests expands geometrically as the tests increase from one cycle to the next.

The ultimate leading edge of testing should be based on formal models such as finite-state machine models, regular expressions, domain specifications, constraint sets, or formal specifications based on the requirements. The tool then automatically generates a covering set of test cases from the model. The advantage is that the focus of writing test cases and testing shifts from designing individual test cases to creating models that faithfully express the requirements.

A more down-to-earth method is to use a capture/playback system to create a base scenario. This forms the base test case, which allows for creation of the variants by simple editing of the base test script. Test automation is an effective methodology for increasing productivity when designing test cases. This is usually well accepted for test execution, but not for test case design. The fact is that the typical manual test case contains 200 keystrokes; the variations between an automated test case and a new test case involve changes in only a dozen keystrokes.

At first glance, it appears easy to automate testing: just get one of the popular test recording/playback tools, record all manual tests and play them back. Unfortunately, life is much more complex and there is more to testing automation then knowing the test tool. Automation testing is also a skill, but very different from the skill required for manual testing. Learning the skill of test automation is an excellent investment into software test engineering that can lead to an interesting and challenging career. The advancement of automated tools has given rise to many new career opportunities for software engineers. Software test automation is still in its infancy stage and the present number of test automation engineers cannot keep pace with the demand.

Test automation can facilitate making testing more efficient than could ever be done by manual testing.
- The most obvious advantage is that it allows running a regressing test quickly on a new version of the program.

- Executing tests frequently. Typically required with iterative projects. This is a clear benefit of automation when many developers are modifying a program and tolerance for long testing cycles is small.
- Allows for more effective use of resources by replacing boring and repetitive tasks [manual testing]. This also stimulates testers to put more effort into designing better test cases.
- Enforces consistency and reputability of tests. A well-designed test will reuse the test data. This is achieved by deleting the test data after the execution of a test, so that the program is ready to be tested the next time with the same or enhanced test data.
- The most obvious motivation for test automation is the earlier time to market by reducing the testing cycle time.

There are a number of problems that can be encountered while trying to automate testing:

Unrealistic expectations: expecting that the new tool will solve all the current testing problems or even that automated testing will replace manual testing.

Poor testing practices like poorly designed test cases, inappropriate documentation etc; automating in any of these instances is not a good idea. Automating chaos will create more chaos faster.

A false sense of security that the automation will find new defects. A test will likely find most defects the first time it runs. Running the same test again is less likely to find any new defects. The fact that the test tool did not find any defects does not mean the program is defect free.

Most importantly, a tool has no imagination, which is the most important attribute of a good tester.

Before doing any testing automation, the pertinent question to ask is: should we automate or not? As per our earlier discussion, the DW application SDLC is an iterative process, which means that the same test cases will be executed many times. Therefore, this is a strong argument for automating DW application testing.

The priority list of what to automate and what not to automate is created based on the following criteria:

Good candidates for automation:
- Short or simple transactions
- The test is executed regularly
- The test is repetitive and reusable
- The test evaluates high risk conditions
- The same test is rerun with many data combinations
- Expected results are stable or easy to generate at runtime
- The test is a baseline test run on several different configurations
- Tasks that are difficult to do manually
- Highest priority features

Poor candidates for automation:
- Long or complex transactions
- One-offs; however, most organizations underestimate the number of times tests will be run. While they think in terms of once or twice, the actual number is more likely to be much bigger over the lifetime of the software.
- Unstable or difficult-to-predict results
- Tests that cross multiple applications

After a decision to automate the test is made, the first important process is to determine the scope of the automation test. This decision is dependent on the project schedule, automation tools, resource skills, and the organization's software testing maturity level and standards. The right test automation framework/scripting technique will help in maintaining the low costs associated with script development and maintenance.

A lot of effort goes into developing and maintaining test automation, and even after it has been built, you may or may not recover the investment. That is why it's very important to perform a good cost/benefit analysis on whatever manual testing you plan to automate. It is very attractive to use test automation in DW Application projects as these are Perpetual Iterative SDLC projects. In the cost/benefit analysis, it may make sense to automate some functions of the DW application rather than invest in full automation development efforts. Just wishing that everything be automated is not a practical strategy.

Here is a misconception that people may have about capture/replay tools. People should not expect to install the capture/replay testing tool, turn on the capture function and begin recording tests that will be used forever and ever. Capturing keystrokes and validating data captured within the script will make the script hard to maintain, besides being useless. In our case study, we demonstrate test scripts that, in fact, are not even created by capture/replay tools.

Determining which tools to use is based on what is being automated and what the purpose of the automation is. In the case study we are presenting in the next section, we are testing a new feed, with high probability that this kind of test will be repeated many times after we have decided to automate the ETL feed into the Data Warehouse. This requires a lot of database querying. This is the same as in our manual and semiautomatic tests, so we will reuse the tool TOAD for this purpose. The other requirement that we must maintain is that all the test results must be stored in the Quality Center [QC]. We could do that manually, but it is time consuming, so we choose to use Quick Test Professional [QTP], as it is well-integrated within the Quality Center. The purpose of the QTP is to run the SQL stored procedure and deposit the results in the QC.

The benefits of combining the QTP and Quality Center:
- Quality Center logs the whole test execution history of QT scripts and traces it back to the business requirements.
- Quality Center can schedule the running time of QTP automation scripts.
- Quality Center can pass variables to QTP automation scripts.
- Quality Center will help QA to manage automation scripts by means of the version control feature.
- Quality Center will bring a user-friendly report to the QAs', as well as to the management team.

The effort of automating an existing manual testing is the same as that of a programmer using a coding language to write programs for automating any other manual process. The scripts need to be maintained over the full life of the DW Application [forever]. The configuration management must apply here too.
The effort of test automation is an investment. Time and resources are needed up front in order to obtain the benefits later on. There is no immediate payoff of automating the current release of the application. The benefit comes from running these automated tests every subsequent release. That is why it becomes important that the scripts can be maintained and kept up to date.

Test automation is like any other software development effort: it's important that those performing the work have the right programming skill sets. A good tester is not necessarily a good fit for doing the test automation. In fact, a good fit is very rare, as the job requirements are quite different than for testing job. That is not to say that testers can't learn to do test automation and be successful; it's just that those two roles are different and the skill sets are different.

Like many other software development efforts, test automation is underestimated, sometimes even grossly underestimated. This is particularly true if the test automation effort is not looked upon as software development, which is usually the case. Test automation is not something that can be done on the side and this should be taken into account when estimating the amount of effort involved.

In our case study, we have developed the stored procedure in manual and semi-automated tests that will be reused in automated tests. This is not done by chance but proceeds from the testing strategy, which should be defined before we start the test automation efforts.

The benefits of automation are pretty obvious. Tests can be run faster, are consistent, and can be run over and over again with less overhead. As more automated tests are added to the test suite, more tests can be run each time thereafter. But manual testing never goes away; instead, the efforts are now focused on more rigorous tests. Automated testing does not replace good test planning, writing test cases or much of the manual testing effort. Test automation is used to test the "error recovery process". This process may run overnight and it creates the report of all the error recovery actions. In the "error recovery process", every validation point is usually exercised using the same common error recovery modules for many validation points.

Automation Risks:

Factors which may affect the automated testing effort and may increase the risk associated with the success of the test include:

Automation Risks	Risk Mitigation
Completion of development of front-end processes	Risk Analysis on Planning Phase
Completion of design and construction of new processes	Automation Testing Standard
Completion of modifications to the local database	Automation Testing Standard
Movement or implementation of the solution to the appropriate testing or production environment	Automation Testing Standard
Stability of the testing or production environment	Automation Testing Standard
Maintaining recording standards and automated processes for the project	
Completion of manual testing through all applicable paths to ensure that reusable automated scripts are valid	

Finally, an objective and convincing picture should be presented to the management via the Automation Test Plan. The Test Plan should present detailed requirements for the success of the test automation. It should include: advantages, disadvantages, pitfalls and risks, human resources, technical resources [hardware and software], and an executions schedule. To gain acceptance of the Test Automation, a presentation to all the stakeholders should be made using the methodology explained in the Test Strategy review section.

Test Automation Case Study

By three methods we may learn wisdom: first, by reflection, which is noblest; second, by imitation, which is easiest; and third by experience; this is the bitterest.

~ Confucius

Background:

A new store is opening at a new location. In order to manage the logistic and marketing information, business team wants to receive all the daily transactions and associated customer information from the store. The new ETL feed job will be created to transfer all the transaction data from the new store to the enterprise DW. The ETL job will be running nightly at midnight. Files including data file and header-trailer information which will be loaded from data warehouse landing area and copy to stage database. A nightly ETL job will load those data from stage database insert into master data warehouse and report database.

Test Case #1

Scope and Description of Data Warehouse of Testing: Verification of Source Data loading into Staging Database.

Objective is to verify data integrity of the transaction files based on the header and trailer information.

Testing assumption: The test script will be reused [company is going to open or relocate other stores], therefore, it is a good candidate for regression and test automation.

Related Data Source [shown on the picture blow fig. 12.1]
Two file are created and placed on the FTP Landing server. Store Business Transaction Data and Header and Trailer files

(1). Incoming Data file [File name: IT_STORETRANSACTIONS.DAT]
(2). Incoming Header and Trailer file for Transaction Data [File name: TT_STORETRANSACTIONS.HT]
(3) Incoming file Transaction Data file is loaded into Stage Database [Oracle 10g database]. Header and Trailer file is used for verification of Transaction Data file.

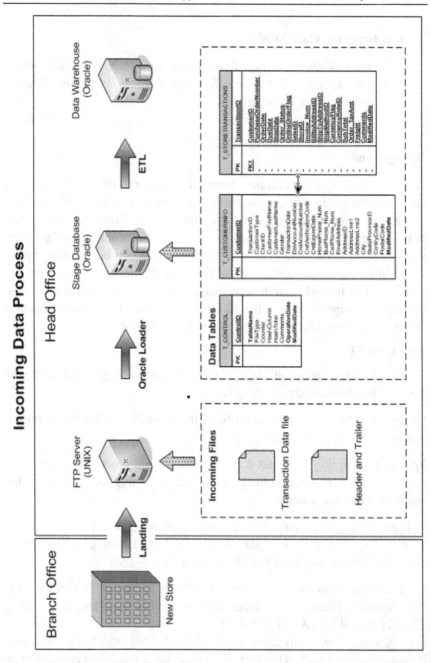

Figure 10.1: The data incoming process

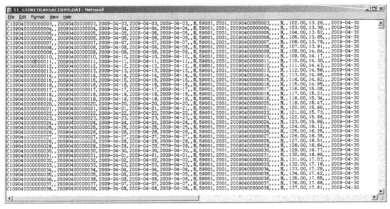

*Figure 10.2: The data file sample for incoming data source
– Store Transaction Feed.*

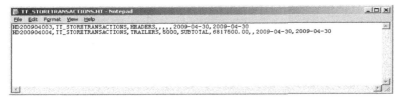

Figure 10.3: The data file sample for incoming data source - Header and Trailer.

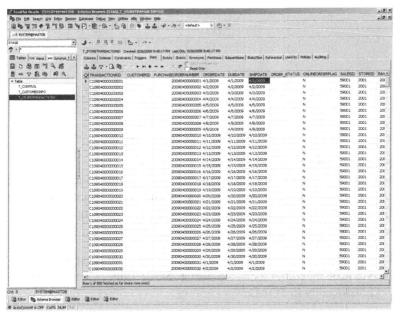

*Figure 12.4: Incoming data populated in the Stage Database
(STAGE.T_STORETRANSACTIONS table).*

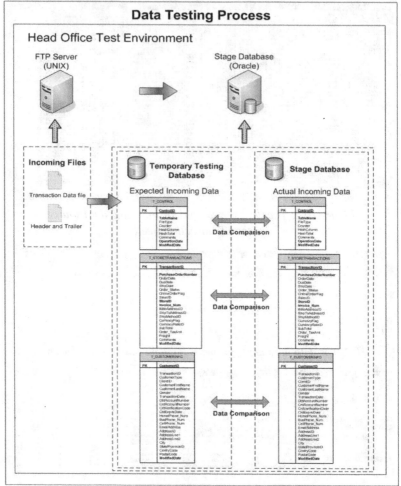

Figure 10.5: Transaction Data Verification Process

Description of the verification process:

Oracle Loader is used to load incoming store Transaction Data file into "TestUser" schema. Test table is created and Store Transaction Data is copied into the test table. In the last step of setting testing environment "test_result" table is created, to store the verification results.

In final verification two test verification methods are presented:
1. Manual [semi-automated] using SQL Table Compare [Ref 34] functionality in TOAD®[1]

---

[1] Registered Trade Marko of Quest Software Company

2. Automated using PL-SQL, HP Quick Test Professional® [test automation tool] and HP Quality Center® [testing management too]

Test Steps:

(1). Reload the data files into TestUser schema by Oracle Loader

```
DROP TABLE testuser.lt_storetransactions CASCADE
CONSTRAINTS;
DROP DIRECTORY test_dir;
CREATE DIRECTORY test_dir AS 'C:\LANDING';
CREATE TABLE testuser.lt_storetransactions
( transactionrid VARCHAR2(50 BYTE),
  customerid VARCHAR2(50 BYTE),
  purchaseordernumber VARCHAR2(50 BYTE),
  orderdate VARCHAR2(50 BYTE),
  duedate VARCHAR2(50 BYTE),
  shipdate VARCHAR2(50 BYTE),
  order_status VARCHAR2(50 BYTE),
  onlineorderflag VARCHAR2(50 BYTE),
  salesid VARCHAR2(50 BYTE),
  storeid VARCHAR2(50 BYTE),
  invoice_num VARCHAR2(50 BYTE),
  billtoaddressid VARCHAR2(50 BYTE),
  shiptoaddressid VARCHAR2(50 BYTE),
  shipmethodid VARCHAR2(50 BYTE),
  currencyflag VARCHAR2(50 BYTE),
  currencyrateid VARCHAR2(50 BYTE),
  subtotal VARCHAR2(50 BYTE),
  order_taxamt VARCHAR2(50 BYTE),
  freight VARCHAR2(50 BYTE),
  comments VARCHAR2(50 BYTE),
  modifieddate VARCHAR2(50 BYTE)
)
ORGANIZATION EXTERNAL
  ( TYPE oracle_loader
    DEFAULT DIRECTORY test_dir
    ACCESS PARAMETERS
      (RECORDS DELIMITED BY NEWLINE
       STRING SIZES ARE IN CHARACTERS
       BADFILE test_dir:'lt_STORETRANSACTIONS.bad'
       LOGFILE test_dir:'lt_STORETRANSACTIONS.log'
       FIELDS TERMINATED BY ','
       OPTIONALLY ENCLOSED BY '"' AND '"'
       MISSING FIELD VALUES ARE NULL
       (transactionrid  CHAR(50),
        customerid  CHAR(50),
        purchaseordernumber  CHAR(50),
        orderdate  CHAR(50),
        duedate  CHAR(50),
        shipdate  CHAR(50),
        order_status  CHAR(50),
        onlineorderflag  CHAR(50),
        salesid  CHAR(50),
        storeid  CHAR(50),
        invoice_num  CHAR(50),
        billtoaddressid  CHAR(50),
        shiptoaddressid  CHAR(50),
```

```
            shipmethodid   CHAR(50),
            currencyflag   CHAR(50),
            currencyrateid   CHAR(50),
            subtotal   CHAR(50),
            order_taxamt   CHAR(50),
            freight   CHAR(50),
            comments   CHAR(50),
            modifieddate   CHAR(50)
             )
           )
        LOCATION (test_dir:'lt_STORETRANSACTIONS.DAT')
     )
   REJECT LIMIT 0
   NOPARALLEL
   NOMONITORING;
```

(2). Copy Transaction data into test table with valid data format.

```
   DROP TABLE testuser.t_storetransactions;
   /
   CREATE TABLE testuser.t_storetransactions
       (transactionrid,
        customerid,
        purchaseordernumber,
        orderdate,
        duedate,
        shipdate,
        order_status,
        onlineorderflag,
        salesid,
        storeid,
        invoice_num,
        billtoaddressid,
        shiptoaddressid,
        shipmethodid,
        currencyflag,
        currencyrateid,
        subtotal,
        order_taxamt,
        freight,
        comments,
        modifieddate
        ) AS
   SELECT
        TRIM(transactionrid),
        TRIM(customerid),
        TO_NUMBER (TRIM(purchaseordernumber))
                AS purchaseordernumber,
        TO_DATE (TRIM(orderdate), 'YYYY-MM-DD')
                AS orderdate,
        TO_DATE (TRIM(duedate), 'YYYY-MM-DD')
                AS duedate,
        TO_DATE (TRIM(shipdate), 'YYYY-MM-DD')
                AS. shipdate,
        TRIM(order_status),
        TRIM(onlineorderflag),
        TRIM(salesid),
```

```
    TRIM(storeid),
    TO_NUMBER (TRIM(invoice_num)) AS invoice_num,
    TRIM(billtoaddressid),
    TRIM(shiptoaddressid),
    TRIM(shipmethodid),
    TRIM(currencyflag),
    TRIM(currencyrateid),
    TO_NUMBER (TRIM(subtotal)) AS subtotal,
    TO_NUMBER (TRIM(order_taxamt)) AS order_taxamt,
    TRIM(freight),
    TRIM(comments),
    TO_DATE (TRIM(modifieddate), 'YYYY-MM-DD')
            AS modifieddate
FROM testuser.lt_storetransactions;
```

(3). Create test result table in TESTUSER schema.

```
DROP TABLE testuser.test_results;
CREATE TABLE testuser.test_results
( item                VARCHAR2(50 BYTE),
  run_version         VARCHAR2(50 BYTE),
  runtime             DATE,
  total_failnmb       NUMBER,
  tablename_source    VARCHAR2(50 BYTE),
  tablename_target    VARCHAR2(50 BYTE),
  test_result         VARCHAR2(50 BYTE),
  comments            VARCHAR2(50 BYTE)
)
TABLESPACE users
PCTUSED     0
PCTFREE     10
INITRANS    1
MAXTRANS    255
STORAGE     (INITIAL         64 k
             MINEXTENTS      1
             MAXEXTENTS      2147483645
             PCTINCREASE     0
             BUFFER_POOL     DEFAULT
            )
LOGGING
NOCOMPRESS
NOCACHE
NOPARALLEL
MONITORING;
```

(4). There are two testing options depending on the testing tools and management decision.

Testing Option I: Manually compare the data tables by using Toad.
(5). Compare Data by using Toad: Open Toad and use the compare data function on Database manual.

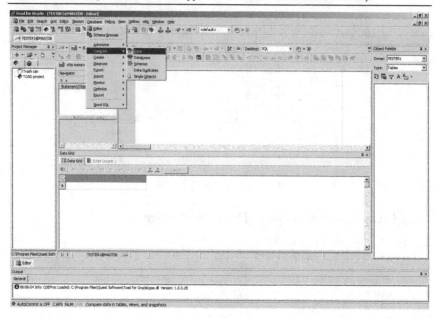

Figure 10.6

Select the table names to compare (STAGE.T_STORETRANSACTION and Table TESTUSER.T_STORETRANSACTION).

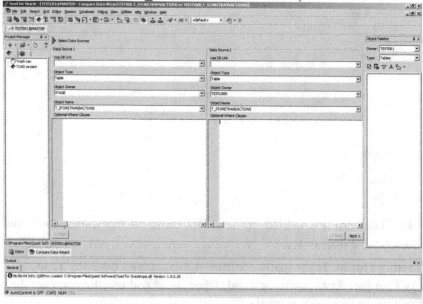

Figure 10.7

Keep the default setup.

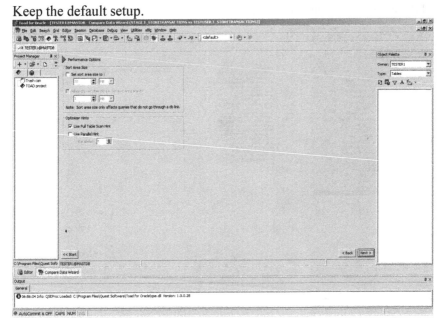

Figure 10.8

All the column names are displayed.

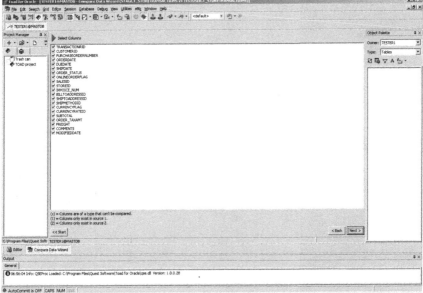

Figure 10.9

Make an order on the test result.

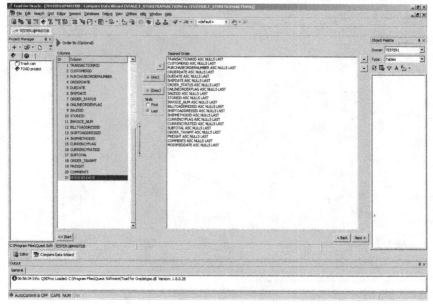

Figure 10.10

Result I: Rows in Table STAGE.T_STORETRANSACTION but not in Table TESTUSER.T_STORETRANSACTION.

Figure 10.11

Tester also can see the query details by click on the "View/Edit Query" button.

```
1   select /*+ FULL(Tbl1) */
2          TRANSACTIONRID, CUSTOMERID, PURCHASEORDERNUMBER, ORDERDATE, DUEDATE
3          , SHIPDATE, ORDER_STATUS, ONLINEORDERFLAG, SALESID, STOREID
4          , INVOICE_NUM, BILLTOADDRESSID, SHIPTOADDRESSID, SHIPMETHODID, CURRENCYFLAG
5          , CURRENCYRATEID, SUBTOTAL, ORDER_TAXAMT, FREIGHT, COMMENTS
6          , MODIFIEDDATE
7   from   STAGE.T_STORETRANSACTIONS Tbl1
8   minus
9   select /*+ FULL(Tbl2) */
10         TRANSACTIONRID, CUSTOMERID, PURCHASEORDERNUMBER, ORDERDATE, DUEDATE
11         , SHIPDATE, ORDER_STATUS, ONLINEORDERFLAG, SALESID, STOREID
12         , INVOICE_NUM, BILLTOADDRESSID, SHIPTOADDRESSID, SHIPMETHODID, CURRENCYFLAG
13         , CURRENCYRATEID, SUBTOTAL, ORDER_TAXAMT, FREIGHT, COMMENTS
14         , MODIFIEDDATE
15  from   TESTUSER.T_STORETRANSACTIONS Tbl2
16  Order by TRANSACTIONRID ASC NULLS LAST, CUSTOMERID ASC NULLS LAST, PURCHASEORDERNUMBER ASC NULLS LAST
17         , ORDERDATE ASC NULLS LAST, DUEDATE ASC NULLS LAST, SHIPDATE ASC NULLS LAST
18         , ORDER_STATUS ASC NULLS LAST, ONLINEORDERFLAG ASC NULLS LAST, SALESID ASC NULLS LAST
19         , STOREID ASC NULLS LAST, INVOICE_NUM ASC NULLS LAST, BILLTOADDRESSID ASC NULLS LAST
20         , SHIPTOADDRESSID ASC NULLS LAST, SHIPMETHODID ASC NULLS LAST, CURRENCYFLAG ASC NULLS LAST
21         , CURRENCYRATEID ASC NULLS LAST, SUBTOTAL ASC NULLS LAST, ORDER_TAXAMT ASC NULLS LAST
22         , FREIGHT ASC NULLS LAST, COMMENTS ASC NULLS LAST, MODIFIEDDATE ASC NULLS LAST
```

Check	Click OK to apply changes	Cancel
		OK

Figure 10.12

Result II: Rows in Table TESTUSER.T_STORETRANSACTION but not in Table STAGE.T_STORETRANSACTION.

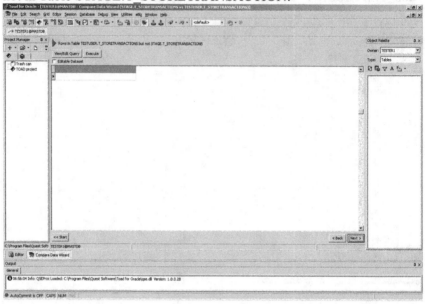

Figure 10.13

Here is the query detail again.

```
select /*+ FULL(Tbl2) */
    TRANSACTIONRID, CUSTOMERID, PURCHASEORDERNUMBER, ORDERDATE, DUEDATE
    , SHIPDATE, ORDER_STATUS, ONLINEORDERFLAG, SALESID, STOREID
    , INVOICE_NUM, BILLTOADDRESSID, SHIPTOADDRESSID, SHIPMETHODID, CURRENCYFLAG
    , CURRENCYRATEID, SUBTOTAL, ORDER_TAXAMT, FREIGHT, COMMENTS
    , MODIFIEDDATE
from    TESTUSER.T_STORETRANSACTIONS Tbl2
minus
select /*+ FULL(Tbl1) */
    TRANSACTIONRID, CUSTOMERID, PURCHASEORDERNUMBER, ORDERDATE, DUEDATE
    , SHIPDATE, ORDER_STATUS, ONLINEORDERFLAG, SALESID, STOREID
    , INVOICE_NUM, BILLTOADDRESSID, SHIPTOADDRESSID, SHIPMETHODID, CURRENCYFLAG
    , CURRENCYRATEID, SUBTOTAL, ORDER_TAXAMT, FREIGHT, COMMENTS
    , MODIFIEDDATE
from    STAGE.T_STORETRANSACTIONS Tbl1
Order by TRANSACTIONRID ASC NULLS LAST, CUSTOMERID ASC NULLS LAST
    , PURCHASEORDERNUMBER ASC NULLS LAST, ORDERDATE ASC NULLS LAST
    , DUEDATE ASC NULLS LAST, SHIPDATE ASC NULLS LAST
    , ORDER_STATUS ASC NULLS LAST, ONLINEORDERFLAG ASC NULLS LAST
    , SALESID ASC NULLS LAST, STOREID ASC NULLS LAST
    , INVOICE_NUM ASC NULLS LAST, BILLTOADDRESSID ASC NULLS LAST
    , SHIPTOADDRESSID ASC NULLS LAST, SHIPMETHODID ASC NULLS LAST
    , CURRENCYFLAG ASC NULLS LAST, CURRENCYRATEID ASC NULLS LAST
    , SUBTOTAL ASC NULLS LAST, ORDER_TAXAMT ASC NULLS LAST
    , FREIGHT ASC NULLS LAST, COMMENTS ASC NULLS LAST
    , MODIFIEDDATE ASC NULLS LAST
```

Figure 10.14

Result III: All the difference between the two tables.

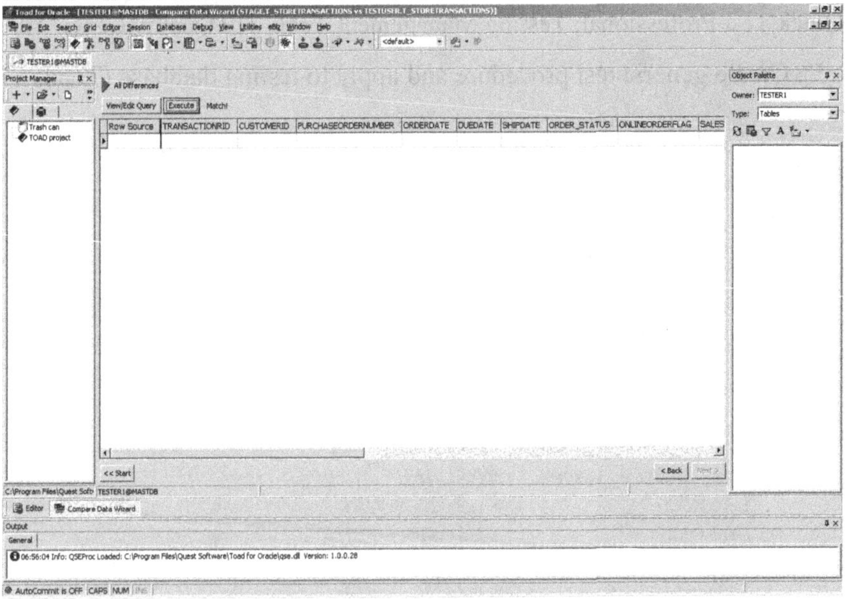

Figure 10.15

The full SQL:

Figure 10.16

Testing Option II: Automation testing script by using PL-SQL and HP Quick Test Professional; Testing Management Tool - HP Quality Center 10.

(5). Create generic test procedure and apply to testing database (Toad)

```
CREATE OR REPLACE Procedure TESTUSER.test002
(local_user IN VARCHAR2, target_user IN VARCHAR2,
tablelike IN VARCHAR2, Run_Version IN VARCHAR2)
IS
    tname VARCHAR2(30);
    CURSOR cur_table
    IS
        SELECT table_name
          FROM all_tables
         WHERE owner = local_user
         AND table_name LIKE tablelike;
    loc_table_name    VARCHAR2 (100);
    tar_table_name    VARCHAR2 (100);
    uatval            PLS_INTEGER;
    sp1val            PLS_INTEGER;
    sqlstatement      VARCHAR2 (2000);
    itemno PLS_INTEGER := 0;
BEGIN
    OPEN cur_table;
    LOOP
        FETCH cur_table INTO tname;
        EXIT WHEN cur_table%NOTFOUND;
        loc_table_name :=  local_user || '.' ||tname;
        tar_table_name :=  target_user || '.' ||tname;
        sqlstatement :=
'SELECT count(*) FROM (SELECT * FROM ' ||
loc_table_name || ' MINUS SELECT * FROM '||
tar_table_name || ')';
        EXECUTE IMMEDIATE sqlstatement
                    INTO uatval;
        sqlstatement :=
'SELECT count(*) FROM (SELECT * FROM ' ||
tar_table_name || ' MINUS SELECT * FROM '||
loc_table_name || ')';
        EXECUTE IMMEDIATE sqlstatement
                    INTO sp1val;
        itemno := itemno + 1;
        IF uatval <> 0 OR sp1val <> 0
        THEN
        Insert into testuser.test_results
( Item, Run_Version, RunTime, Total_FailNMB,
TableName_Source, TableName_Target, Test_Result,
Comments)
        Values ( itemno, Run_Version, SYSTIMESTAMP,
uatval + sp1val, loc_table_name, tar_table_name,
```

```
'Fail', itemno || '.' ||loc_table_name || ' has
different result!');
      ELSE
      Insert into testuser.test_results
( Item, Run_Version, RunTime, Total_FailNMB,
TableName_Source, TableName_Target, Test_Result,
Comments)
Values ( itemno, Run_Version, SYSTIMESTAMP, uatval +
sp1val, loc_table_name, tar_table_name, 'Pass',
null);
      END IF;
   END LOOP;
   CLOSE cur_table;
   COMMIT;
END;
```

Figure 10.17: The test procedure is applied to database

(6). Execute the Test Procedure to compare the data in Staging database by HP Quick Test Professional 10 (VB Script).

```
P1= Parameter( "ProcedureName")
P2= Parameter( "Target")
P3= Parameter( "Actual")
P4= Parameter( "CheckPt")
P5= Parameter( "ReleaseNo")

RunStoredProcedure P1, P2, P3, P4, P5
```

```
VerifyTestResult   P5

Function RunStoredProcedure(ProcedureName,
Parameter1, Parameter2, Parameter3, Parameter4 )
          Dim Cmd
        DatabaseName="MASTDB"
        UserID="TESTUSER"
        Pwd="TEST1234"

        '  Create the database object
        Set Cmd = CreateObject("ADODB.Command")

        ' Activate the connection.
        Cmd.ActiveConnection = "DRIVER={Microsoft
ODBC for Oracle}; SERVER=" & DatabaseName & ";User
ID=" & UserID & ";Password=" & Pwd & " ;"

        ' Set the command type to Stored Procedures
        Cmd.CommandType = 4
        Cmd.CommandText = ProcedureName

        ' Define Parameters for the stored procedure
        Cmd.Parameters.Refresh

        'Pass Parameters to Stored Procedure
        Cmd.Parameters(0).Value = Parameter1
        Cmd.Parameters(1).Value = Parameter2
        Cmd.Parameters(2).Value = Parameter3
        Cmd.Parameters(3).Value = Parameter4

        ' Execute the stored procedure
        Cmd.Execute()
        Set Cmd = Nothing
End Function

Function VerifyTestResult (Parameter4 )
         Dim Rs, Conn
         Dim SQL
        DatabaseName="MASTDB"
        UserID="TESTUSER"
        Pwd="TEST1234"
        '  Create the database object
        Set Conn=Createobject("adodb.connection")
        Set  Rs=Createobject("adodb.recordset")
        Conn.open="Driver={Microsoft ODBC for
Oracle};Server= " & DatabaseName &_
          ";Uid="& UserID &";Pwd="& Pwd &";"

        SQL="select TEST_RESULT, TABLENAME_SOURCE,
COMMENTS  from TESTUSER.TEST_RESULTS where
RUN_VERSION = '"& Parameter4 &"' "
        Rs.open SQL,Conn,3,3

        Do While Not Rs.eof
                Results = Rs("TEST_RESULT")
```

```
SourceTable = Rs("TABLENAME_SOURCE")
If Results = "Pass" Then
            Reporter.ReportEvent micPass,
Parameter4 & "_Table Comparison", SourceTable
        Else
            Reporter.ReportEvent micfail,
Parameter4 & "_Table Comparison", SourceTable
        End if
        Rs.movenext
    Loop
    Rs.close
    Set Rs=nothing
    Conn.close
    Set Conn=nothing
End Function
```

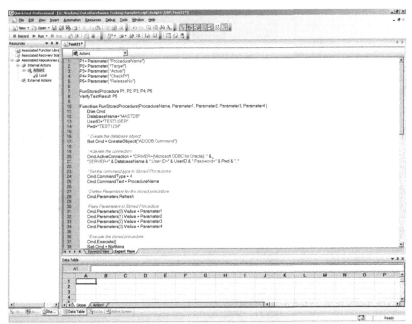

Figure 10.18: HP Quack Test Professional Script

(7). Run QTP script and save the test results into Quality Center

Note:

Before connecting HP Quick Test Professional to HP Quality Center, make sure the add-ins are installed in Quality Center:

- HP Quality Center Connectivity Add-in
- Quick Test Professional Add-in

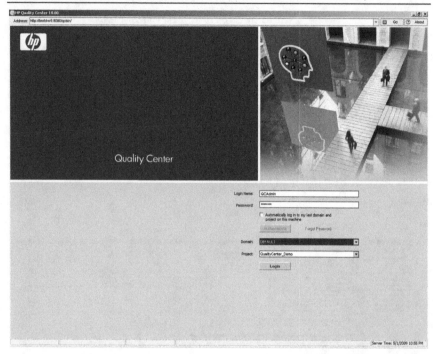

Figure 10.19: Execute the Quack Test Professional Script through HP Quality Center 10

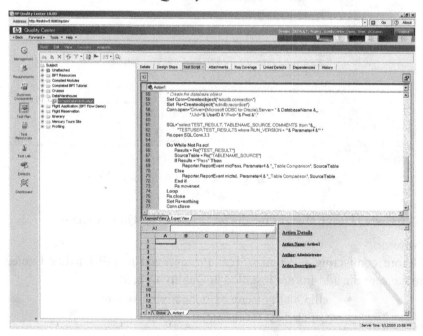

Figure 10.20: Automation Test Script in Test Plan

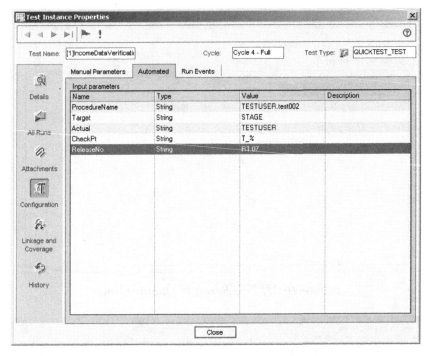

Figure 10.21: Update the script parameters in Quality Center

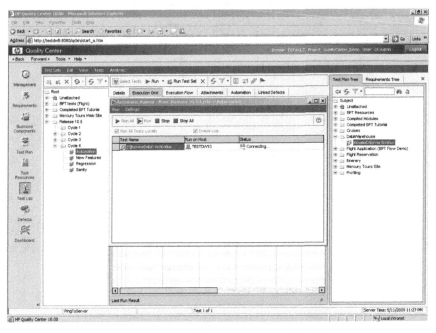

Figure 10.22: Run the script in Quality Center

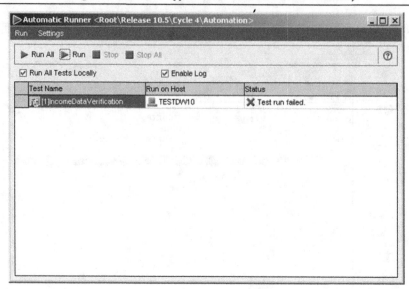

Figure 10.23: Test Result in Quality Center

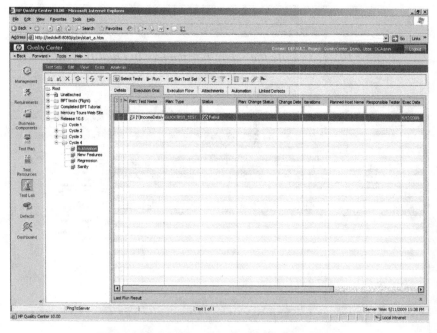

Figure 10.24: Test Result in Quality Center

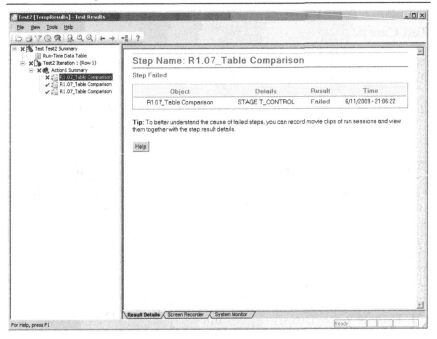

Figure 10.25: Test Result - Failed

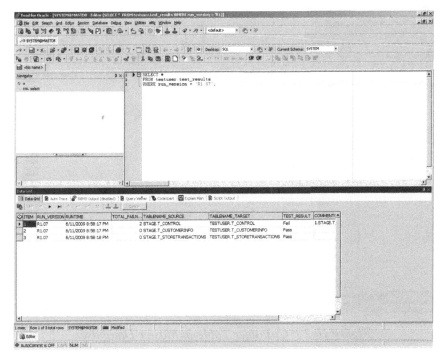

Figure 10.26: The same Test Result in TOAD

Test Case #2

<u>Scope and Description of Data Warehouse of Testing</u>: ETL verification [bases on business rules] in Master Database.

<u>Introduction</u>: There is an ETL job from Staging Database load the customer information data into Data Warehouse.
It includes some data format changes below.

	Column Name	Staging Database		Master Database	
		Format	Data Sample	Format	Data Sample
1	CUSTOMERID	Varchar2 .(14byte)	CU1010000000 77	Number (10)	1000000077
2	CUSTOMERNAME	Different column for First and Last name	John Smith	First and Last Name in a same column	John Smith
3	STATUS	None	Null	Active with Transactions	Active
4	CRDACCOUNTNUMBER	Varchar2 (20byte)	C109040000000 016	Number (15)	1090400000000 16

Related Data Source: Oracle 10g Database

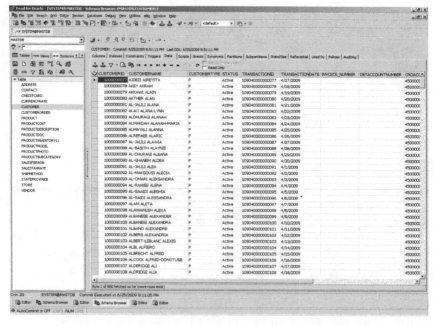

Figure 10.27: Customer Table in Master Database.

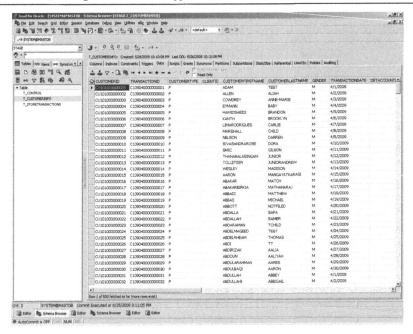

Figure 10.28: Customer Table in Staging Database.

Test Option I: Compare the data by using SQL script (Toad)

```
SELECT COUNT (*) AS Record_Count,
        DECODE(SUM(DECODE(mast.customerid,
stag.customerid, 0 , 1)), 0, 'Pass', 'Fail') AS
customerid,
        DECODE(SUM(DECODE(mast.customername,
stag.customername, 0 , 1)), 0, 'Pass', 'Fail') AS
customername,
        DECODE(SUM(DECODE(mast.customertype,
stag.customertype, 0 , 1)), 0, 'Pass', 'Fail') AS
customertype,
        DECODE(SUM(DECODE(mast.status, stag.status, 0
, 1)), 0, 'Pass', 'Fail') AS status,
        DECODE(SUM(DECODE(mast.transactionid,
stag.transactionid, 0 , 1)), 0, 'Pass', 'Fail') AS
transactionid,
        DECODE(SUM(DECODE(mast.transactiondate,
stag.transactiondate, 0 , 1)), 0, 'Pass', 'Fail') AS
transactiondate,
        DECODE(SUM(DECODE(mast.invoice_number,
stag.invoice_number, 0 , 1)), 0, 'Pass', 'Fail') AS
invoice_number,
        DECODE(SUM(DECODE(mast.dbtaccountnumber,
stag.dbtaccountnumber, 0 , 1)), 0, 'Pass', 'Fail') AS
dbtaccountnumber,
        DECODE(SUM(DECODE(mast.crdaccountnumber,
stag.crdaccountnumber, 0 , 1)), 0, 'Pass', 'Fail') AS
crdaccountnumber,
        DECODE(SUM(DECODE(mast.storeid, stag.storeid,
.0 , 1)), 0, 'Pass', 'Fail') AS storeid,
```

```
      DECODE(SUM(DECODE(mast.addressid,
stag.addressid, 0 , 1)), 0, 'Pass', 'Fail') AS
addressid,
      DECODE(SUM(DECODE(mast.modifieddate,
stag.modifieddate, 0 , 1)), 0, 'Pass', 'Fail') AS
modifieddate,
      DECODE(SUM(DECODE(mast.salesid, stag.salesid,
0 , 1)), 0, 'Pass', 'Fail') AS salesid
  FROM mastdb.customer mast,
      (SELECT TO_NUMBER (SUBSTR (customerid, -10))
AS customerid, (customerfirstname || ' ' ||
customerlastname) AS customername,
            customertype,
            'Active' AS status,
            TO_NUMBER (SUBSTR (transactionid, -
15)) AS transactionid,
            transactiondate,
            NULL AS invoice_number,
            TO_CHAR (dbtaccountnumber) AS
dbtaccountnumber,
            TO_CHAR (crdaccountnumber) AS
crdaccountnumber,
            NULL AS storeid,
            addressid,
            modifieddate,
            NULL AS salesid
      FROM stage.t_customerinfo) stag
  WHERE mast.customerid = stag.customerid
    AND mast.modifieddate = stag.modifieddate;
```

Figure 10.29: Test Script and results.

Test Option II: Compare the data by using SQL script insides VB script [Descriptive Programming]; the test result will be report to a Microsoft Excel data sheet.

The SQL Query:

```
SELECT COUNT (*) AS Record_Count,
       SUM(DECODE(mast.customerid, stag.customerid, 0
, 1)) AS customerid,
       SUM(DECODE(mast.customername,
stag.customername, 0 , 1)) AS customername,
       SUM(DECODE(mast.customertype,
stag.customertype, 0 , 1)) AS customertype,
       SUM(DECODE(mast.status, stag.status, 0 , 1))
AS status,
       SUM(DECODE(mast.transactionid,
stag.transactionid, 0 , 1)) AS transactionid,
       SUM(DECODE(mast.transactiondate,
stag.transactiondate, 0 , 1)) AS transactiondate,
       SUM(DECODE(mast.invoice_number,
stag.invoice_number, 0 , 1)) AS invoice_number,
       SUM(DECODE(mast.dbtaccountnumber,
stag.dbtaccountnumber, 0 , 1)) AS dbtaccountnumber,
       SUM(DECODE(mast.crdaccountnumber,
stag.crdaccountnumber, 0 , 1)) AS crdaccountnumber,
       SUM(DECODE(mast.storeid, stag.storeid, 0 , 1))
AS storeid,
       SUM(DECODE(mast.addressid, stag.addressid, 0 ,
1)) AS addressid,
       SUM(DECODE(mast.modifieddate,
stag.modifieddate, 0 , 1)) AS modifieddate,
       SUM(DECODE(mast.salesid, stag.salesid, 0 , 1))
AS salesid
  FROM mastdb.customer mast,
       (SELECT TO_NUMBER (SUBSTR (customerid, -10))
AS customerid,
       (customerfirstname || ' ' || customerlastname)
AS customername,
                customertype,
                'Active' AS status,
                TO_NUMBER (SUBSTR (transactionid, -
15)) AS transactionid,
                transactiondate,
                NULL AS invoice_number,
       TO_CHAR (dbtaccountnumber) AS dbtaccountnumber,
       TO_CHAR (crdaccountnumber) AS crdaccountnumber,
                NULL AS storeid,
                addressid,
                modifieddate,
                NULL AS salesid
       FROM stage.t_customerinfo) stag
  WHERE mast.customerid = stag.customerid
   AND mast.modifieddate = stag.modifieddate;
```

The details of VB Script:

```
'=========================================================
' VBScript Source File -- Database Testing
```

```vbscript
' Version: 1.0
' Feature and Description: ETL Verification
' Author:
' Date:
'=========================================================
Dim DBName, UserID, Pwd, TableName, SQL
Dim objXL, objWb, objP1, objP2 ' Excel object
variables

' Open Excel data sheet
Set objXL = CreateObject ("Excel.Application")
objXL.Visible = true ' Show Excel window
objXL.DisplayAlerts = False 'Disable Excel Alert
Set objWb = objXL.WorkBooks.Open(GetPath +
"DBTest.xls")
Set objP1 =
objXL.ActiveWorkBook.WorkSheets("General_Info")
Set objP2 =
objXL.ActiveWorkBook.WorkSheets("Test_Results")
objP1.Activate  ' Show General_Info Page

' Retrieve Database configuration information from
the Excel file
DBName = Trim(objP1.Cells(12, 3).Value)
UserID = Trim(objP1.Cells(13, 3).Value)
Pwd = Trim(objP1.Cells(14, 3).Value)
TableName = Trim(objP1.Cells(15, 3).Value)
SQL = Trim(objP1.Cells(16, 3).Value)
SaveAsName = Trim(objP1.Cells(8, 3).Value)

' Apply default value
If CStr(SaveAsName) = "" Then
   SaveAsName = "Test_Results"
End If

' Manually enter User ID
If UserID = "" Then
   UserID = Inputbox ("Please enter your Database
User ID:")
   UserID = Trim (UserID)
End If

' Manually enter Password
If Pwd = "" Then
   Pwd = Inputbox ("Please enter your Database
Password:")
   Pwd = Trim (Pwd)
End If

objP2.Activate  ' Show Test_Results Page
' Create the database object and connect to database
Dim oCon: Set oCon =
WScript.CreateObject("ADODB.Connection")
Dim oRs: Set oRs =
WScript.CreateObject("ADODB.Recordset")
Dim ConnectInf
If oCon is Nothing Then
```

```
        Msgbox "Database Connection Error"
        Wscript.Quit
End If
ConnectInf = "dsn=" & DBName & "; uid="& UserID &";
pwd="& Pwd &";"
oCon.Open ConnectInf
'   Execute the query
Set oRs = oCon.Execute(SQL)
If Err.number <> 0 Then
        Msgbox "SQL Error!!"
        Wscript.Quit
End If

' Read all query records and insert the results into
excel file
Do until oRs.EOF
    For i = 0 to oRs.Fields.Count-1 'Get the Field
Count
            objP2.Cells(i+2, 1).Value = "'" & CStr(i+1)
            objP2.Cells(i+2, 2).Value = "'" & TableName
            'Get Field Name
            objP2.Cells(i+2, 3).Value = "'" &
oRs.Fields(i).Name
            'Get Field Value
            objP2.Cells(i+2, 4).Value = "'" &
oRs.Fields(i).Value
            ' Check the test results without the row of
record count
      If UCase(oRs.Fields(i).Name) <> "RECORD_COUNT" Then
            If oRs.Fields(i).Value = 0 Then
                ' There is no different record
                objP2.Cells(i+2, 5).Value = "'Pass"
            Else
                ' There is different record
                objP2.Cells(i+2, 5).Value = "'Fail"
                objP2.Cells(i+2, 5).Font.ColorIndex = 3
'Make color red
                objP2.Cells(i+2, 5).Font.Bold = True
'Make font bold
            End If
        End If
    Next
    oRs.MoveNext
Loop

oRs.Close
oCon.Close
Set oRs = Nothing
Set oCon = Nothing

' Save results
objWb.SaveAs GetPath + SaveAsName + ".xls"

objXL.DisplayAlerts = True
objXL.Quit
Set objP1 = Nothing
```

```
Set objP2 = Nothing
Set objWb = Nothing
Set objXL = Nothing

Function GetPath()
    ' Retrieve the script path
    DIM path
    path = WScript.ScriptFullName ' Script name
    GetPath = Left(path, InstrRev(path, "\"))
End Function
```

The test steps:

Step #1: Create an Excel test template following the format below:

Page 1 – "General_Info" and insert the Query inside the SQL cell:

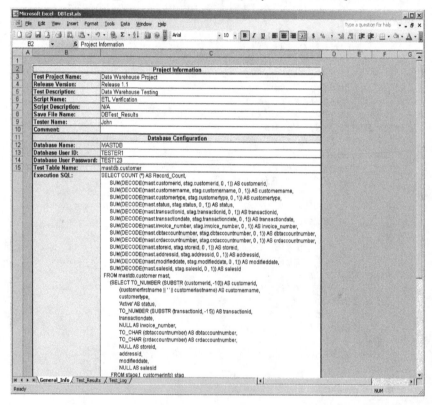

Figure 10.30

Data Sheet Page 2 – "Test_Results":

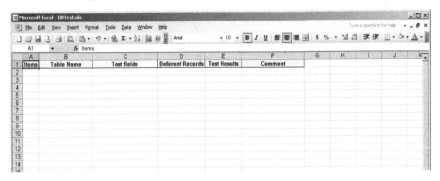

Figure 10.31

Step #2: Copy the VB Script into a Notepad and save the file with an extension name "vbs". Make sure this vbs file and the Excel template are located in a same folder. Here our example file name is Test_Script_V1.0.vbs and DBTest.xls.

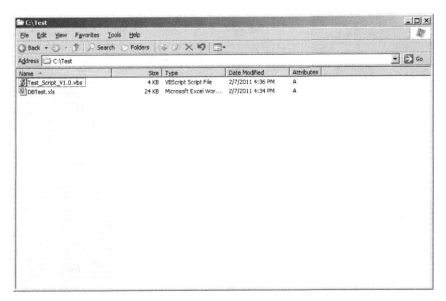

Figure 10.32

Step #3: Double click on the Test_Script_V1.0.vbs file to run it. The test result file is generated.

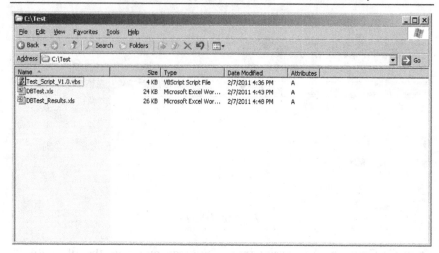

Figure 10.33

Step #4: Check the execution report on the result file.

	A	B	C	D	E	F	G
1	Items	Table Name	Test fields	Deferent Records	Test Results	Comment	
2	1	mastdb.customer	RECORD_COUNT	5000			
3	2	mastdb.customer	CUSTOMERID	0	Pass		
4	3	mastdb.customer	CUSTOMERNAME	0	Pass		
5	4	mastdb.customer	CUSTOMERTYPE	0	Pass		
6	5	mastdb.customer	STATUS	5000	Fail		
7	6	mastdb.customer	TRANSACTIONID	0	Pass		
8	7	mastdb.customer	TRANSACTIONDATE	0	Pass		
9	8	mastdb.customer	INVOICE_NUMBER	0	Pass		
10	9	mastdb.customer	DBTACCOUNTNUMBER	0	Pass		
11	10	mastdb.customer	CRDACCOUNTNUMBER	0	Pass		
12	11	mastdb.customer	STOREID	0	Pass		
13	12	mastdb.customer	ADDRESSID	1	Fail		
14	13	mastdb.customer	MODIFIEDDATE	0	Pass		
15	14	mastdb.customer	SALESID	0	Pass		

Figure 10.34

Test Case #3

<u>Scope and Description of Data Warehouse of Testing</u>: Landing files verification by using UNIX Shell Script.

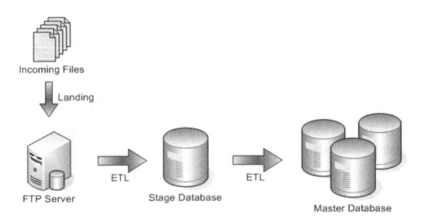

Figure 10.35

There are two files will be loaded into landing area (FTP server) every day. Here are the file names and directory:

FTP directory: /var/ftp/Landing/
Header/Trailer file: transaction.ht

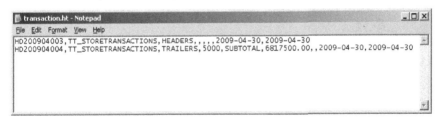

Figure 10.36: Header/Trailer file

Data file: customerinfo.csv

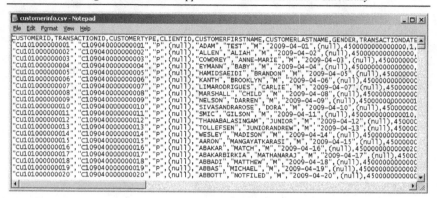

Figure 10.37: Data file - customerinfo.csv

This Bash Shell script will check the file existence, the loading date and the data creation date.

```
#!/bin/bash
# This script gives information about a file.

LoadDate="$(stat -c %x
/var/ftp/Landing/transaction.ht | awk '{print $1}')"
DateInFile="$(cut -f8 -d","
/var/ftp/Landing/customerinfo.csv | head -n 1)"
TodayDate="$(date +%Y-%m-%d)"
YesterdayDate="$(date -d yesterday +%Y-%m-%d)"

# Check the file existence
if [ -f /var/ftp/Landing/transaction.ht ]; then
    echo "Header and Trailer file not exist..."
    exit 1
fi
if [ -f /var/ftp/Landing/customerinfo.csv ]; then
    echo "Transaction data not exist..."
    exit 1
fi

#Check the file landing date
if [ "$LoadDate" == "$TodayDate" ]
then
echo "The file is loaded today."
else
echo "The file is NOT loaded today."
# Mail to tester
mail -s "The loading date is not match..."
tester@abc.com
fi

#Check the date inside file
if [ "$DateInFile" == "$YesterdayDate" ]
then
echo "The data is generated yesterday."
else
echo "The data is NOT generated yesterday."
# Mail to tester
```

```
mail -s "The creation date is not match..."
tester@abc.com
fi
```

The shell execution:

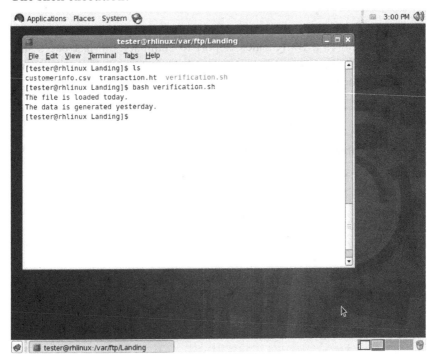

Figure 10.38

Test Case #4

Scope and Description of Data Warehouse of Testing:
Report Testing (VBScript Automation)

Testing Background:
Business requires a Daily Sales Report to summary the last day's sale information for the stores. Test team needs to create an automation script to compare the report data in data warehouse.

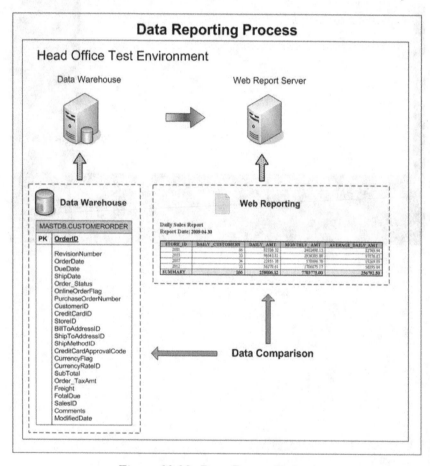

Figure 10.39: Data Report Process

The data in the "mastdb.customerorder" table:

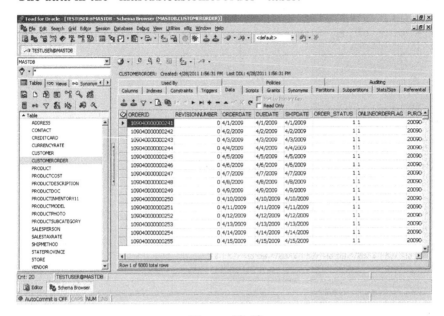

Figure 10.40

Sample Report: Daily Sales Report

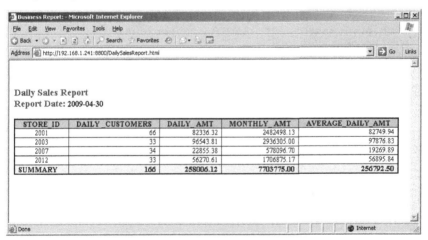

Figure 10.41: Daily Sales Report

The Automation Testing Script (VB Script):

```
Option Explicit
Public objIE 'Create object of IE Application
Dim OutputText

'Get the database access information
Dim SID, DBName, UserID, Password
SID = "MASTDB"
DBName = "MASTDB"

'Enter the output file name
Dim ConnectInf, Query
Dim OutputInfo, OutputFile
OutputFile = "TestResults.txt"
OutputInfo = ""

'Enter Database User ID
UserID = Inputbox ("Please enter your User ID:")
UserID = Trim (UserID)

'Enter Database Password
Password = Inputbox ("Please enter your Password:")
Password = Trim (Password)

'---------- Open IE -------------
'Navigate the report webpage.
NavWeb("http://192.168.1.241:8800/")
'Click the link: Daily Sales Report
WebLink "Daily Sales Report"

'Create the database connection
Dim oCon: Set oCon =
WScript.CreateObject("ADODB.Connection")
Public oRs: Set oRs =
WScript.CreateObject("ADODB.Recordset")

If oCon is Nothing then
     Msgbox "Connection error!"
End if

ConnectInf = "Provider=OraOLEDB.Oracle;Data Source="&_
            SID"; User ID="& UserID &"; Password="&_
            Password &";"
oCon.Open ConnectInf

'Create the testing output file with a time stamp.
CreateOutput OutputFile, vbCrLf & Cstr(Now) & " -------
------------" & vbCrLf

'---------------- Query I -----------------
'Get the last date from database.
Query = "SELECT TO_CHAR (MAX (orderdate), 'YYYY-MM-DD')
AS orderdate FROM mastdb.customerorder"
'Execute the query
SQLExecution Query

'Compare the Report Date value on the report web page.
```

```
If oRs.Fields(0).Value = ReadInfo("Report Date:", 10)
Then
    OutputInfo = "The value of Report Date is correct!"
Else
    OutputInfo = "Database has a different report date:"
                    & oRs.Fields(0).Value
End If

'Write the result into test output file.
CreateOutput OutputFile, OutputInfo

'--------------- Query II ----------------
'Get the daily summery from database.
Query = "SELECT storeid AS store_id, COUNT(customerid)
    AS daily_customers, SUM (totalpurchase) AS
    daily_amt  FROM (SELECT customerid, storeid, SUM
    (subtotal + order_taxamt) AS totalpurchase FROM
    mastdb.customerorder WHERE orderdate = (SELECT
    MAX(orderdate) FROM mastdb.customerorder) GROUP BY
    storeid, customerid) GROUP BY storeid ORDER BY
    storeid"

'Execute the query
SQLExecution Query

Dim ColumnNbr, RowNbr
Dim oF
Dim TotalCustomers, TotalDailyAmt, TotalMonthlyAmt,
AvgDailyAmt
OutputInfo = ""
RowNbr = 1
TotalCustomers = 0
TotalDailyAmt = 0
TotalMonthlyAmt = 0
AvgDailyAmt = 0
OutputInfo = ""

'Compare the value on the report web page.
'This part will check the first three columns:
STORE_ID, DAILY _CUSTOMERS, and DAILY_ AMT
'Loop for different rows
While Not oRs.EOF
    ColumnNbr = 0
    'Loop for different columns
    For Each oF in oRs.Fields
        'Compare the query result and the value on web
report
        If Round(oRs.Fields(ColumnNbr).Value, 2) <>
CDbl(WebTable(RowNbr+1, ColumnNbr+1)) Then
            OutputInfo = OutputInfo & "The database has a
different value for cell (" & CStr(RowNbr + 1) & ", " &
CStr(ColumnNbr + 1)  & "): "oRs.Fields(ColumnNbr).Value
& vbCrLf
        End If
        Select Case ColumnNbr
          'Sum the DAILY_CUSTOMERS Summary
          Case 1
```

```
                TotalCustomers = TotalCustomers +
Round(oRs.Fields(ColumnNbr).Value, 2)
          'Sum the DAILY_AMT Summary
          Case 2
              TotalDailyAmt = TotalDailyAmt +
Round(oRs.Fields(ColumnNbr).Value, 3)
        End Select
        ColumnNbr = ColumnNbr + 1
    Next
    oRs.MoveNext
    RowNbr = RowNbr + 1
Wend


'Compare the query result and the value on web report
for DAILY_CUSTOMERS Summary
If TotalCustomers <> CDbl(WebTable(RowNbr + 1, 2)) Then
    OutputInfo=OutputInfo & "The database has a
different DAILY_CUSTOMERS Summary value: " &
TotalCustomers & vbCrLf
End If
'Compare the query result and the value on web report
for DAILY_AMT Summary
If TotalDailyAmt <> CDbl(WebTable(RowNbr + 1, 3)) Then
    OutputInfo = OutputInfo & "The database has a
different DAILY_AMT Summary value: "TotalDailyAmt &
vbCrLf
End If

''---------------- Query III ----------------
''Check how many days in the current month
Query = "SELECT COUNT (1) FROM (SELECT mcoo.orderdate
    FROM (SELECT * FROM mastdb.customerorder mco,
    (SELECT MAX(orderdate) AS maxorderdate FROM
    mastdb.customerorder) ldt WHERE TO_CHAR
    (mco.orderdate, 'Year') = TO_CHAR (ldt.maxorderdate,
    'Year') AND TO_CHAR (mco.orderdate, 'Month') =
    TO_CHAR (ldt.maxorderdate, 'Month')) mcoo GROUP BY
    mcoo.orderdate)"

'Execute the query
SQLExecution Query

Dim DateNbr
DateNbr = CInt(oRs.Fields(0).Value)

''---------------- Query IV ----------------
''Get the Monthly Amount
Query = "Select mcoo.storeid, SUM (mcoo.subtotal +
    mcoo.order_taxamt) AS totalpurchase FROM (SELECT *
    FROM mastdb.customerorder mco, (SELECT MAX
    (orderdate) AS maxorderdate FROM
    mastdb.customerorder) ldt WHERE TO_CHAR
    (mco.orderdate, 'Year') = TO_CHAR (ldt.maxorderdate,
    'Year') AND TO_CHAR (mco.orderdate, 'Month') =
    TO_CHAR (ldt.maxorderdate, 'Month')) mcoo GROUP BY
    mcoo.storeid   ORDER BY storeid"

'Execute the query
```

```
SQLExecution Query

Dim AverageAmt
RowNbr = 1


'Compare the value on the report web page.
'This part will check the last two columns: MONTHLY_AMT
and AVERAGE_DAILY_AMT
'Loop for different rows
While Not oRs.EOF
    If Round(oRs.Fields(1).Value, 2) <>
CDbl(WebTable(RowNbr + 1, 4)) Then
    OutputInfo=OutputInfo & "The database has a
different value for cell (" & CStr(RowNbr+1) & ", 4): "
& oRs.Fields(1).Value & vbCrLf
    End If
    AverageAmt =
Round((CDbl(oRs.Fields(1).Value)/DateNbr), 2)

    If AverageAmt <> CDbl(WebTable(RowNbr + 1, 5)) Then
    OutputInfo=OutputInfo & "The database has a
different value for cell (" & CStr(RowNbr+1) &", 5): "
& CStr(AverageAmt) & vbCrLf
    End If

    TotalMonthlyAmt = TotalMonthlyAmt +
Round(oRs.Fields(1).Value, 2)

    oRs.MoveNext
    RowNbr = RowNbr + 1
Wend

'Compare the query result and the value on web report
for MONTHLY_AMT Summary
If TotalMonthlyAmt <> CDbl(WebTable(RowNbr + 1, 4))
Then
    OutputInfo = OutputInfo & "The database has a
different MONTHLY_AMT Summary value: " & TotalCustomers
& vbCrLf
End If

'Compare the query result and the value on web report
for AVERAGE_DAILY_AMT Summary
AvgDailyAmt = Round((TotalMonthlyAmt / DateNbr), 2)
If AvgDailyAmt <> CDbl(WebTable(RowNbr + 1, 5)) Then
    OutputInfo=OutputInfo & "The database has a
different AVERAGE_DAILY_AMT Summary value: " &
AvgDailyAmt & vbCrLf
End If

'Problem output
If OutputInfo = "" Then
    CreateOutput OutputFile, "The entire test data is
correct!"
Else
    CreateOutput OutputFile, "Failed on the Data
Verification: " & vbCrLf & OutputInfo
End If
```

```
Msgbox "Script Finish!"

Err.Clear
Set oRs = Nothing
Set oCon = Nothing
Set objIE = Nothing

'================= Functions =================
'SQL Query Execution
Function SQLExecution(SQL)
    'SQL is the Query details
    Set oRs = oCon.Execute(Query)
    If Err.number <> 0 Then
        Msgbox "SQL Query Failed !"
        Wscript.Quit
    End If
    If oRs.EOF Then
        Msgbox "No query result!"
        Wscript.Quit
    End If
End Function


'Read infomation from the innerText on web page.
Function ReadInfo(ReadBefore, ReadSize)
    'ReadBefore is the Text before reading area.
    'ReadSize is the reading character number.
    'Sample: ReadInfo("Report Date:", 10)
    Dim FullText
    Dim StartPort
    FullText = objIE.Document.body.innerText
    StartPort = (InStr(LCase(FullText),
LCase(ReadBefore)) + Len(ReadBefore))
    ReadInfo = Trim(Right(FullText, Len(FullText) -
StartPort))
    ReadInfo = Left(ReadInfo, ReadSize)
End Function


'Read infomation from the table on web page.
Function WebTable(RowID, ColumnID)
    'RowID is the row number on the table.
    'ColumnID is the column number on the table.
    'Sample: WebTable(2, 1)
    Dim Table
    Table = objIE.Document.getElementsByTagName("table")
    WebTable = Table.rows(RowID - 1).cells(ColumnID -
1).innerText
End Function


'Check the IE activity status
Function IEActive()
    Do While objIE.Busy Or (objIE.READYSTATE <> 4)
        Wscript.Sleep 80
    Loop
    WScript.Sleep 80
End Function


'Navigate Website
```

```
Function NavWeb(WebAddress)
   Set objIE =
CreateObject("InternetExplorer.Application")
   objIE.Visible = True
   objIE.Navigate WebAddress
   IEActive
End Function

'Click the Link on web page
Function WebLink(NavLink)
   Dim i
   Dim TotalLinks
   Dim Linkexist
   Linkexist = False
   TotalLinks = objIE.Document.Links.Length
   For i=0 To TotalLinks-1
      If
InStr(LCase(objIE.Document.Links(i).innerHTML),
LCase(NavLink)) > 0 Then
                objIE.Document.Links(i).click
                Linkexist = True
                Exit For
      End If
   Next
   If Linkexist = False Then
      Wscript.Echo "Script could not find this link: "
& NavLink
      Wscript.Quit
   Else
      IEActive
   End If
End Function

'Get the current working directory
Function GetPath()
   ' Retrieve the script path
   DIM path
   path = WScript.ScriptFullName ' Script name
   GetPath = Left(path, InstrRev(path, "\"))
End Function

'Create Output txt file
Function CreateOutput(Filename, FileContent)
   Dim fso, OutputFile, MyFile
   OutputFile = GetPath + Filename
   Set fso = CreateObject("Scripting.FileSystemObject")
      If fso.FileExists(OutputFile) = False Then
      Set MyFile = fso.CreateTextFile(OutputFile,
True)
      MyFile.Close
      end if
   Set MyFile = fso.OpenTextFile(OutputFile, 8, True)
   MyFile.WriteLine(FileContent)
   MyFile.Close
   Set fso = Nothing
   Set MyFile = Nothing
End Function
```

There are total 4 SQL queries included in the script.

Query I: Get the last date from database which we are going to use it compare the Report Date.

```
SELECT TO_CHAR (MAX (orderdate), 'YYYY-MM-DD')
          AS orderdate
FROM mastdb.customerorder;
```

ORDERDATE

2009-04-30

Query II: Get the daily summery from database.

```
SELECT     storeid AS store_id,
           COUNT (customerid) AS daily_customers,
           SUM (totalpurchase) AS daily_amt
   FROM (SELECT    customerid, storeid,
           SUM (subtotal + order_taxamt) AS totalpurchase
           FROM mastdb.customerorder'
           WHERE orderdate = (SELECT MAX (orderdate)
                              FROM mastdb.customerorder)
           -- Make sure customer is counted only once
           GROUP BY storeid, customerid)
GROUP BY storeid
ORDER BY storeid;
```

STORE_ID	DAILY_CUSTOMERS	DAILY_AMT
2001	66	82336.32
2003	33	96543.81
2007	34	22855.38
2012	33	56270.61

Query III: Check how many days in the current month.

```
SELECT COUNT (1)
   FROM (SELECT    mcoo.orderdate
           FROM (SELECT *
                   FROM mastdb.customerorder mco,
                   (SELECT MAX (orderdate) AS maxorderdate
                      FROM mastdb.customerorder) ldt
                   WHERE TO_CHAR (mco.orderdate, 'Year') =
                       TO_CHAR (ldt.maxorderdate, 'Year')
                   AND TO_CHAR (mco.orderdate, 'Month') =
                       TO_CHAR (ldt.maxorderdate, 'Month')) mcoo
       GROUP BY mcoo.orderdate);
```

COUNT(1)

30

Query IV: Get the Monthly Summary Amount.

```
SELECT   mcoo.storeid,
         SUM (mcoo.subtotal + mcoo.order_taxamt)
             AS totalpurchase
  FROM (SELECT * FROM mastdb.customerorder mco,
            (SELECT MAX (orderdate) AS maxorderdate
                FROM mastdb.customerorder) ldt
         WHERE TO_CHAR (mco.orderdate, 'Year') =
               TO_CHAR (ldt.maxorderdate, 'Year')
           AND TO_CHAR (mco.orderdate, 'Month') =
               TO_CHAR (ldt.maxorderdate, 'Month')) mcoo
GROUP BY mcoo.storeid
ORDER BY storeid;
```

STOREID	TOTALPURCHASE
2001	2482498.13
2003	2936305
2007	578096.7
2012	1706875.17

Testing Folder:
The automation script is saved into a txt file and name it as *.vbs.

Before test execution:

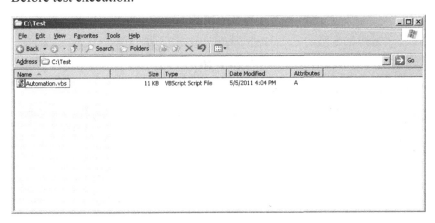

Figure 10.42

After test execution:

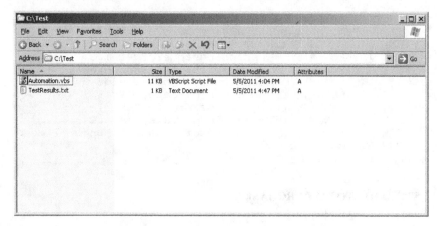

Figure 10.43

Test Result file: All passed

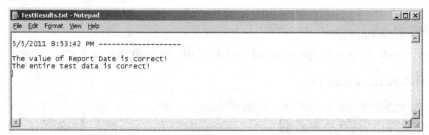

Figure 10.44

or Test Result file: failed on the verification

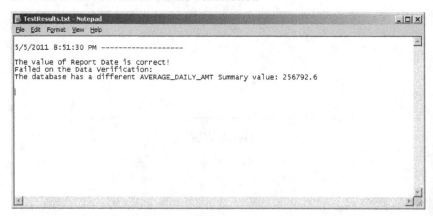

Figure 10.45

Test Management

Test management is a part of the QA process. It is the process of organizing and controlling all the testing activity and testing objects required for the overall testing efforts. Basic tools used for test management may include:

- Word processors
- Spreadsheets
- Automated test management tools

Test management activities may be broken down into different tasks: organization, planning, test case design, test execution, and reporting. Testing efforts are as effective as they are able to convey the testing status with some measurement of the quality of the project.

Test organization is a test management activity. It requires organizing and maintaining testing objects and activities. It includes organizing the test team, including the relationship with other teams. The typical test objects to be managed are:

- Test scripts [manual and automated]
- Test data [acquiring and maintaining]
- Test software tools
- Test hardware environments

Test planning is the overall set of tasks that address the questions of why, what, where, and when to test. The reason why a given test is created is called a test motivator [for example, a specific requirement must be verified]. What should be tested is broken down into many test cases for a project. Where to test is answered by determining and documenting the needed software and hardware configurations. When to test is resolved by tracking iterations [or cycles, or time period] to the testing.

Test case design is the process of capturing the specific steps required to complete a given test. This addresses the question of how something will be tested and is where somewhat abstract test cases are developed into more detailed test steps, which in turn will become test scripts [either manual or automated].

Test execution entails running the tests by assembling sequences of test scripts into a suite of tests. This is a continuation of answering the question of how something will be tested [more specifically, how the testing will be conducted].

Test reporting is how the various results of the testing effort are analyzed and communicated. This is used to determine the current status of project testing, as well as the overall level of quality of the application or system.

The testing effort will produce a great deal of information about the quality of the product being tested. This information can be extracted to populate **quality metrics** that can be used to define, measure, and track quality goals of projects. The information can further be used for future project planning and for tracking the continuous quality improvement process.

There are a number of test tools that can be used in the defect management process, such as: HP QualityCenter®, IBM Rational ClearQuest TestManager®, etc. The previous section – "Case Study: HP Quality Center" has been used to demonstrate the use of automated tools in the Test Management process.

Defect Tracking

A very common type of data produced by testing, one which is often a basic source of quality metrics, is *defects*. Defect status is not static, but changes over time. In addition, multiple defects are often related to one another. Effective defect tracking is crucial to both testing and development teams.

HP Quality Center, a defect-tracking tool, is used in the case study to track defects. After the defect life cycle is implemented in the QC, the defect management process is intuitive:

Once a defect is identified, a ticket is created and a **new** defect is assigned to the development team manager for further analysis. Upon determining that it is a genuine problem, the gatekeeper [development team leader] determines who should be "working" the defect, and the defect is **assigned** to that developer. The defect resolution could be a clarification and/or may require actual modifications to the ETL process, for the defect resolution.

If a breakdown in communication occurs during the tracking of a defect, the verification team could inadvertently fail scripts based on incorrect expected results. For example, a defect is logged regarding bad data. A decision is made to modify the ETL process and to scrap the data. The verification team needs to be notified of this change, and to subsequently modify the expected results within the appropriate test scripts.

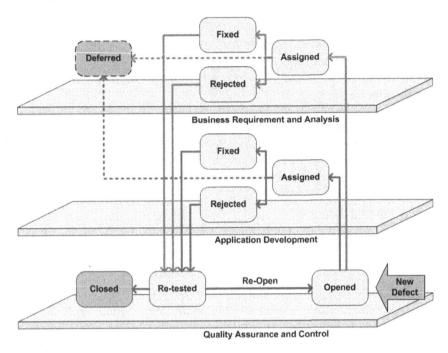

Figure 11.1 Defect Tracking

References:

1. Ralph Kimball, Margy Ross: *The Data Warehouse Toolkit: The Complete Guide to Dimensional Modeling* – published by Wiley & Sons

2. Dean Leffingwell & Don Widrig. 2003. *Managing Software Requirements A use case Approach*

3. Gause Donald, Gerald Weinberg. 1989. *Exploring Requirements: Quality before Design*. New York: Dorset House Publishing

4. INMON, W.H.: '*Building the data warehouse*' John Wiley and Sons, September 1998,New York, 1997]

5. Alan Cooper: *The Inmates Are Running the Asylum*

6. Edward Tufte: *The Visual Display of Quantitative Information, 2nd edition*

7. Stephen Few: *Show Me the Numbers: Designing Tables and Graphs to Enlighten*

8. Stephen Few: *Information Dashboard Design: The Effective Visual Communication of Data*

9. Dr. William S. Cleveland: *The Elements of Graphing Data*

10. Steven D. Levitt, Stephen J. Dubner: *Freakonomics*

11. Donald A. Norman: *Things That Make Us Smart: Defending Human Attributes In The Age Of The Machi*

12. Dr. John Maeda: *The Laws of Simplicity [Simplicity: Design, Technology, Business, Life]*

13. Marty Neumeier: *The Brand Gap: Expanded Edition*

14. Daniel H. Pink: *A Whole New Mind: Why Right-Brainers Will Rule the Future*

15. Dr. Jacques Bertin: *Semiology of Graphics: Diagrams, Networks, Maps*

16. Dr. Colin Ware: *Information Visualization, Second Edition: Perception for Design* [Interactive Technologies]

17. Chip Heath, Dan Heath: Switch: *How to Change Things When Change Is Hard*

18. Chip Heath: Made to Stick: *Why Some Ideas Survive and Others Die*

19. Card, Mackinlay, and Shneiderman: *Readings in Information Visualization*

20. Dr. Ian Ayres: *Super Crunchers*

21. Toby Segaran and Jeff Hammerbacher: Beautiful Data: *The Stories Behind Elegant Data Solutions*

22. David Levy: *Tools of Critical Thinking*

23. Dr. Betty Edwards: *Drawing on the Right Side of the Brain*

24. Dr. Betty Edwards: *Drawing on the Artist Within*

25. Tony Schwartz: *What Really Matters: Searching for Wisdom in America*

26. Michael Michalko: *Thinker Toys* [2nd Edition]

27. Kelsey Ruger: *The Owner's Manual for the Brain*

28. Dr. Boris. Beizer: *Software Testing Techniques*

29. Chic Thompson: *What A Great Idea*

30. Peter Drucker: *The Age of Discontinuity*

31. Plato: *The Republic* [Written 360 B.C.E]

32. Hand, David, Heikki Mannila, and Padhraic Smyth: *Principles of Data Mining*

33. Dr. Andrew Abela: *The Presentation: A Story About Communicating Successfully With Very Few Slides*

34. Dr. Viktor Frankl: *Man's Search for Meaning*

35. Dr. Howard Wainer: *Visual Revelations: Graphical Tales of Fate and Deception from Napoleon Bonaparte to Ross Perot*

36. Dr. Howard Wainer.: *Graphic Discovery: A Trout in the Milk and Other Visual Adventures*

37. Dr. Howard Wainer: *Picturing the Uncertain World: How to Understand, Communicate and Control Uncertainty through Graphical Display*

38. Dr. Wilkinson, L.: *The Grammar of Graphics*

39. Dr. Boris Beizer: *Software System Testing and Quality Assurance*

40. Dr. Mihaly Csikszentmihalyi: *Flow: The Psychology of Optimal Experience*

41. Watts Humphrey: *Managing the Software Process*

42. **Dr. Deming's 14 Points:**

1. Constancy of purpose

Create constancy of purpose for continual improvement of products and service to society, allocating resources to provide for long range needs rather than only short term profitability, with a plan to become competitive, to stay in business, and to provide jobs

2. The new philosophy

Adopt the new philosophy. We are in a new economic age, created in Japan. We can no longer live with commonly accepted levels of delays, mistakes, defective materials, and defective workmanship. Transformation of Western management style is necessary to halt the continued decline of business and industry.

3. Cease dependence on mass inspection

Eliminate the need for mass inspection as the way of life to achieve quality by building quality into the product in the first place. Require statistical evidence of built in quality in both manufacturing and purchasing functions.

4. End lowest tender contracts

End the practice of awarding business solely on the basis of price tag. Instead require meaningful measures of quality along with price. Reduce the number of suppliers for the same item by eliminating those that do not qualify with statistical and other evidence of quality. The aim is to minimize total cost, not merely initial cost, by minimizing variation. This may be achieved by moving toward a single supplier for any one item, on a long term relationship of loyalty and trust. Purchasing managers have a new job, and must learn it

5. Improve every process

Improve constantly and forever every process for planning, production, and service. Search continually for problems in order to improve every activity in the company, to improve quality and productivity, and thus to constantly decrease costs. Institute innovation and constant improvement of product, service, and process. It is management's job to work continually on the system (design, incoming materials, maintenance, improvement of machines, supervision, training, retraining).

6. Institute training on the job
Institute modern methods of training on the job for all, including management, to make better use of every employee. New skills are required to keep up with changes in materials, methods, product and service design, machinery, techniques, and service

7. Institute leadership
Adopt and institute leadership aimed at helping people do a better job. The responsibility of managers and supervisors must be changed from sheer numbers to quality. Improvement of quality will automatically improve productivity. Management must ensure that immediate action is taken on reports of inherited defects, maintenance requirements, poor tools, fuzzy operational definitions, and all conditions detrimental to quality

8. Drive out fear
Encourage effective two way communication and other means to drive out fear throughout the organization so that everybody may work effectively and more productively for the company

9. Break down barriers
Break down barriers between departments and staff areas. People in different areas, such as Leasing, Maintenance, Administration, must work in teams to tackle problems that may be encountered with products or service.

10. Eliminate exhortations
Eliminate the use of slogans, posters and exhortations for the work force, demanding Zero Defects and new levels of productivity, without providing methods. Such exhortations only create adversarial relationships; the bulk of the causes of low quality and low productivity belong to the system, and thus lie beyond the power of the work force.

11. Eliminate arbitrary numerical targets
Eliminate work standards that prescribe quotas for the work force and numerical goals for people in management. Substitute aids and helpful leadership in order to achieve continual improvement of quality and productivity.

12. Permit pride of workmanship
Remove the barriers that rob hourly workers, and people in management, of their right to pride of workmanship. This implies, among other things, abolition of the annual merit rating (appraisal of performance) and of Management by Objective. Again, the responsibility of managers, supervisors, foremen must be changed from sheer numbers to quality.

13. Encourage education

Institute a vigorous program of education, and encourage self improvement for everyone. What an organization needs is not just good people; it needs people that are improving with education. Advances in competitive position will have their roots in knowledge.

14. Top management commitment and action

Clearly define top management's permanent commitment to ever improving quality and productivity, and their obligation to implement all of these principles. Indeed, it is not enough that top management commit themselves for life to quality and productivity. They must know what it is that they are committed to—that is, what they must do. Create a structure in top management that will push every day on the preceding 13 Points, and take action in order to accomplish the transformation. Support is not enough: action is required!

43. An example, comparing tables A and B, and the primary key of both tables is ID:

```
SELECT MIN(TableName) as TableName, ID, COL1,
       COL2, COL3 ...
FROM
( SELECT 'Table A' as TableName, A.ID, A.COL1,
         A.COL2, A.COL3, ...
    FROM A
    UNION ALL
  SELECT 'Table B' as TableName, B.ID, B.COL1,
         B.COl2, B.COL3, ...
    FROM B
) tmp
GROUP BY ID, COL1, COL2, COL3 ...
HAVING COUNT(*) = 1
ORDER BY ID
```

The above query returns all rows in either table that do not completely match all columns in the other. In addition, it returns all rows in either table that do not exist in the other table. It handles nulls as well, since GROUP BY normally consolidates NULL values together in the same group. If both tables match completely, no rows are returned at all.